W9-BQU-119

BASIC GENETICS
A HUMAN APPROACH

Third Edition

BSCS

KENDALL/HUNT PUBLISHING COMPANY
4050 Westmark Drive Dubuque, Iowa 52002

AUTHORS

Joseph D. McInerney, BSCS, revision coordinator
Ronald G. Davidson, MD, Toronto Hospital for Sick Children, Canada
Edward Drexler, Pius XI High School, Milwaukee, Wisconsin

BSCS PRODUCTION STAFF

Dee A. Miller, project secretary
Cathrine M. Monson, editor

BSCS ADMINISTRATIVE STAFF

Timothy H. Goldsmith, PhD, chairman, board of directors
Joseph D. McInerney, director
Michael J. Dougherty, PhD, assistant director
Lynda B. Micikas, PhD, assistant director
Larry Satkowiak, chief financial officer

AUTHORS OF PREVIOUS EDITIONS

First Edition	Second Edition
Donna Day Baird, PhD	Edward Drexler
Barton Childs, MD	Joseph D. McInerney
Ronald G. Davidson, MD	Jeffrey C. Murray, MD
Edward Drexler	
Bennie Latimer	
John M. Opitz, MD	
Stuart F. Spicker, PhD	
Ted K. Tsumura	

Cover Photos: Image of DNA strands and laboratory photo © 1998 PhotoDisc. Images of children and background cover art courtesy of Corel.

Copyright © 1983, 1991, 1999 by the BSCS

Library of Congress Catalog Card Number: 97-76030

ISBN 0-7872-3023-5

The first edition of this work was supported by the Development in Science Education (DISE) Program of the National Science Foundation, Grant No. SED-7918983. However, the opinions expressed herein do not necessarily reflect the position or policy of the National Science Foundation, and no official endorsement should be inferred. "Living with Cystic Fibrosis" was developed under a grant from the Cystic Fibrosis Foundation, Rockville, Maryland. However, the opinions expressed herein do not necessarily reflect the position or policy of the Cystic Fibrosis Foundation, and no official endorsement should be inferred.

Printed in the United States of America
10 9 8 7 6 5 4 3 2 1

CONTENTS

GETTING STARTED

ON THE COVER

The cover of *Basic Genetics: A Human Approach* shows what this magazine is all about: people. People come in all sizes and shapes. They are of different colors and of different ages. They have different interests and different ideas. Genetics is, in fact, the study of the causes of variation in living things.

But people also have some things in common; for example, all have a genetic heritage. Each person is unique. But each person also is a product of evolution. Each of us carries hereditary material that has been transmitted from generation to generation over millions of years.

This magazine is about people. But specifically, this magazine is about you, about your parents and relatives, and about the children you might have someday. In part, you are unique because of the genes you inherited from your mother and father. That genetic legacy has been and continues to be acted on by your environment and your experiences. An important aim of education is to have you understand yourself—your capacities and your limitations. Those are, in part, genetically determined. This magazine will pro-

vide new insights into yourself as a product of past heredity, present capability, and future potential.

Basic Genetics: A Human Approach presents genetics as a discipline that will help you understand those inherited instructions that in part make you what you are. Genetics has been taught most often using organisms other than humans. But examples from *Homo sapiens* can illustrate the principles of genetics as well as any other organism. In addition, human examples provide the benefits of applied biology—the ways in which research in biology affects individuals, families, and society. You will come to understand what a pedigree is and how genetics helps to determine certain human characteristics. The laws of probability and their effect on you and on the children you might have can be important for your future and the future of your family.

Not all human inheritance is positive. People inherit characteristics for normal human structures and functions, but they also can inherit abnormalities. Genetically determined disorders are frequently serious, because they last longer than the infectious diseases we combat so often. Even genetic disorders, however, can sometimes be anticipated. In some

cases, we can prevent the disorder or make it less severe. But the medical interest in genetic disorders and diseases leads quickly to social, moral, and legal dilemmas.

We are living at a time when new genetic and medical knowledge provides power. Along with that power comes new responsibility. Simply because we know *how* to do something does not mean that we *ought* to do it. The rapid growth of information and technology means that all of us will incur responsibility for decisions that previous generations never had to face. The range of choices human beings must make grows ever wider.

We can and should use our new knowledge to benefit individuals and society. But how do we determine those benefits? The study of human genetics leads to many issues that are outside the realm of scientific inquiry. Furthermore, the study of human genetics has a practical, applied value, but this value may well be different for each person. Perhaps no two individuals, after studying these materials, will come to exactly the same conclusions. Each of us is different, and our backgrounds make us feel differently about certain issues.

This particular area of biology is of great importance to each of us. Knowledge of genetics will give you a better understanding of how to cope with what is, while giving you an understanding of your capacities to change what might be.

We are eager to hear about your use of these materials and their effect on you. Please send your comments and suggestions to the BSCS at the address below.

Timothy H. Goldsmith, PhD
Chairman, BSCS Board of
    Directors
Yale University
New Haven, Connecticut

Joseph D. McInerney, Director
Biological Sciences Curriculum
    Study (BSCS)
5415 Mark Dabling Blvd.
Colorado Springs, Colorado
    80918-3842

LOOKING AHEAD

It often is true that learning is easier and more enjoyable if we can relate new information to ourselves. All of us need to see some relationship between new facts and our own lives or the lives of our families and friends. Because this entire magazine is devoted to human genetics, it probably is wise to take a moment to think about what is included. Consider how the new ideas you will encounter here relate to things you already have learned.

Most people have heard something about genetics. Usually, genetics is presented in the context of plants and of animals other than humans. Most science books emphasize the genetics of pea plants and fruit flies and sometimes other living things, such as cows, mice, or certain flowering plants.

This magazine approaches genetics differently. The big ideas are similar, but the organism that we examine here most often is *Homo sapiens*. Remember that much of our present knowledge of genetics came from the study of plants and nonhuman animals. But it also is important to understand that the same genetic principles apply broadly to all living things.

Furthermore, our knowledge of the genetics of humans is growing rapidly. No one should be surprised if some of the things scientists now think are correct turn out to be incorrect later on. Indeed, many of things that were included in the 1991 edition of this magazine had to be revised or omitted for this edition, because they were no longer correct. That is the way science works.

Anyone who studies this magazine for any time at all should master some of the basic principles of genetics. But don't stop there. Nearly every section of this magazine opens the door to more extensive examination of the genetic principles and of the ethical, legal, and social issues that arise from new knowledge and technology in genetics. Those latter issues are unique to humans, because humans are the only organisms that can anticipate the long-term consequences of choices made today. The future will demand that you—as an individual and as a member of society—make some complex and difficult choices. The authors of this magazine hope that your experience in thinking through such decisions now will help you when you face difficult choices later on.

A MAGAZINE, NOT A BOOK

Basic Genetics: A Human Approach is a magazine, not a book. By magazine, we do not mean that it comes out every week or every month. Instead, it has the *form* of a magazine. It includes articles, stories, editorials, interviews, and other features. Although this magazine makes sense if you read it from front to back, you do not have to use it that way. You can skip around. You also might find that you will spend more time with some parts than others.

This magazine, however, is like a book in one way. The authors had certain goals in mind when they wrote it. They settled on some things they thought everyone should gain from reading the stories and articles that are included here. To get an idea of what is in this magazine and to assess your own prior knowledge, complete the inventory provided by your teacher. Rate yourself before you read anything in this magazine. Rate yourself again after you have completed your studies. Your score should improve dramatically.

Why Study Human Genetics?

Hardly a day goes by when there isn't a newspaper article or a television report on a newly discovered gene. The media frequently will predict what this discovery might mean to individuals and to society. The Human Genome Project is filling computers with genetic information from the intensive research going on in the United States and around the world. Again, the media attempt to interpret this information and to indicate what the impact might be on you, your community, your country, or the world. Do you want to learn how genetics affects human life only from the press and television?

We feel strongly that your science classes will provide accurate information that you can use to sort sense from nonsense so that you will have the basis for intelligent and thoughtful decisions about genetic policies, both public and personal. Three sets of reasons help us understand why we all should know something about genetics: (1) philosophical reasons, (2) reasons of health, and (3) social reasons.

Philosophical Reasons

One definition of philosophy is a body of rules governing a field of study, in this case human genetics. The powers of Nature are great; we are part of nature and we need to know what its laws are, how we must obey them, and what the limits are to our capacity to adapt ourselves to that which can't be changed.

Life was well developed before *Homo sapiens* ever appeared; the biological rules were already laid down by millions upon millions of years of evolution. Each human cell is under the control of its genes. Biologists have written books describing that control. Our genes help to determine the rates at which all reactions go, and whether they go at all. They determine the size, shape, and specificity of the proteins that make up organelles, cells, organs, and organ systems, as well as which functions are reserved to which cells and organs. They help to determine not only what we are, but what we *can* and what we *cannot* be.

The embodiment of those rules is the genetic material—the DNA. DNA has two principal properties. We've mentioned one already, its power to specify the limits within which cells, organs, and individuals can work. The other is the power of replication by which cells (and individuals) reproduce and maintain the continuity of living organisms.

The power of reproduction. The DNA code is simple; it's a language of four letters that can be arranged to encode biological meaning. It can replicate itself, leading to the production of new cells, some of which are gametes capable of uniting to produce new individuals. All of the potential properties of such new individuals derive from the specificity of the inherited DNA, so each must share the biological qualities of the parents.

Mutations that occur in human DNA lead to variations in the expression of traits. An additional source of variation is provided by the union of gametes containing chromosomes derived from two parents, so that genetic differences can be transmitted from both sides of the family. And even beyond that, further variation is derived from the power of chromosomes to recombine and so to endow each new person with unique combinations of genes. So, the power of self-replication, together with mutational changes and chromosomal mixing, accounts for the ability of the DNA to specify both sameness and difference, conformity, and variation. Each person is unique, possessing properties all his own or her own.

The continuity of life. The universality of DNA is evidence that all forms of life are related, sharing similar mechanisms of control over life processes and variability. We recognize this kinship by considering that if all organisms share these same processes of life, there must be something in common about the ways living organisms have adapted themselves to the Earth; there must be some rules that all organisms obey.

So, these properties of the genetic material—its control over cellular organization and metabolism, its power of replication (including its capacity to change by mutation), and its central position throughout the whole of living organisms—have profound implications for our understanding of human personal individuality, the identity of humans as a species, our place in the biosphere, and the interdependence of all forms of life.

REASONS OF HEALTH

The second set of reasons for being interested in human genetics has to do with health. Let's begin by returning to the role of the genes in specifying the molecules that determine the structure, organization, and function of the cells, and recall that in doing so they set limits on the capacity of those cells (as well as organs, organ systems, and individuals) to respond to the conditions of the environment. Now, because mutations and chromosomal recombination are the sources of genetic variability, and therefore of individuality, each of us has his or her own private ways of responding to experiences, of defending biological integrity, or of succumbing to threats that overwhelm those defenses. The inability to maintain homeostasis (the process of keeping internal conditions constant despite changes in the external environment) in the face of adversity is called disease, and we will see that disease is simply a result of a bad fit between the competence of the mechanism to maintain a steady state and forces that oppose it.

The essence of that idea is that this fit, or misfit, is an individual matter; experiences that are catastrophic for some are of no particular concern for others. Everything depends on the harmony, or lack thereof, of genes and experiences. Of course, some mutant genes exert effects so disruptive that they represent a bad fit with any environment normally encountered in life, and the diseases they promote are usually represented as genetic disorders. But we're learning about the genetic contribution to many other disorders, until recently thought to be due only to agents of the environment: infections, high blood pressure, heart attacks, stomach ulcers, cancers, and pretty well everything else. Apart from infections and nutritional deficiencies (which also are genetically influenced), cures aren't so easily carried out; treatment is the rule, and that isn't always so successful either. So, we are turning to prevention; never to have a disease, even a curable one, certainly produces less wear and tear on the system. Because prevention involves those who wish to avoid becoming patients, the key to that desirable state is knowledge of one's genetic constitution.

In general, we wish for others what is useful for ourselves, and nowhere is that desire more valid than in the family. When we can avert genetic damage by some timely intervention, parents are usually prepared to avail themselves of it. An important addition to the ranks of health-care professionals is that of the genetic counselor who is sought out by parents to help them understand the odds that their unborn will inherit a genetic disorder, as well as the appropriate treatment if that happens.

Disease is an intensely personal experience. When sick, we feel isolated, dependent, and threatened. We are unlikely to think of ourselves as a small and not very significant element in a mighty evolutionary program. But disease is evidence of natural selection at work and is comprehensible only in that context. The end result of this natural selection is death. The death rate is highest *in utero* (before birth) and in old age. There is evidence that at least one-half of conceptions do not reach term. Most of the causes of this very early death are diseases that are genetic; half of miscarriages have chromosome abnormalities, and most of the rest have one or another of a wide range of abnormalities. We know that perhaps five percent of newborn infants have some sort of congenital abnormality and that malformations are the principal cause of death during the first year. We can understand that high death rate only in terms of selection; reproduction is chancy, susceptible to many errors, and fetuses that would be unlikely in any case to survive in the extrauterine world die *in utero*.

There is a sudden increase in the death rate of infants shortly after birth. These are the inborn errors of metabolism (genetically determined biochemical disorders). They have their onset usually after birth because the mother's circulation has fulfilled the metabolic needs that the newborn infant cannot manage on its own because of some genetic deficiency.

The healthiest years are those of childhood, adolescence, and young adult life. The incidence of genetic disorders declines during this and later phases of life; indeed, accidents are the principal cause of death and infections the main cause of illness during those years. The death rate begins to rise in the thirties and forties and stretches out into old age. Now there are very few so-called "genetic" diseases, but many of the disorders that cause premature death are familial and more and more are being shown to be associated with genetic risk factors. So, genes are necessary, but not sufficient of themselves to cause most of the diseases of later life. Personal habits, living conditions, and chance events are required to bring genetic susceptibility to overt disease.

So it appears that natural selection is at work throughout life eliminating those whose genotypes do not fit well with the conditions under which individuals are compelled to live, or choose to live. The worst fits appear *in utero* and are aborted spontaneously; the best fit is observed in adult life. One supposes that the genes that figure in the death of the very old must be those that set the upper limit for the life span of any human being. Another way of putting this is that the species is at its most variable genetically during intrauterine life, and at its least variable in very old age.

The genes involved in the diseases of adult life appear to be compatible with most environments, emerging as damaging only under fairly specific conditions. We should direct research in prevention to discovery of people at risk and to helping them to design a life that skirts the genetic traps that lurk within them. The genes of the diseases of the unborn or the newly born are more broadly incongruent, often leading to maximum developmental problems and death in any conceivable circumstance. The certainty of such unstoppable outcomes has led to the paradox of preventing disease by preventing the birth of affected persons. One medical justification for that form of prevention is that it is in tune with nature, which has brought about the elimination of most of such abnormal fetuses anyway.

No one needs reminding that such a logic does violence to conventional mores both within the medical profession and society at large. Which leads to the third set of reasons for studying human genetics, those having to do with the social impact of this understanding of the most intimate details of people's inherited makeup.

SOCIAL REASONS

Fears aroused by the possibility of misuse of genetic information have attracted much more attention than the knowledge itself. Often people who are ignorant of the meaning, let alone of the facts, of genetics have pronounced resoundingly on how it should be used—and not used. Still, there is no question that this new knowledge threatens values, and public debate is useful, healthy, and necessary.

What is threatened? Privacy and confidentiality to start with. When we learn that someone has a particular gene, Mendelian laws assign probabilities that relatives have it, too. Suppose the gene can be harmful. Is there an obligation to tell others of their risk? We can make a good case for doing so in connection with some diseases at least, but to do so may breach the confidential relationship between patient and doctor. Suppose a woman who has a hemophilic son and can be shown to be a carrier refuses to allow anyone to reveal her genotype to her sisters. Don't the latter have the right to information that could help them in their own reproductive lives? The doctor could inform them in opposition to the mother's wishes, but in so doing he betrays a confidence. A dilemma!

How about public screening for genes that contribute to disease? Mothers are sometimes unaware that their newborn babies are being tested for such disorders and are shocked by a telephone call urging a second, confirmatory test. A test for what? Why? What's happening to my baby? Wouldn't it be better if people know the rationale of screening, what it is for, and what to expect?

Some industrial companies are testing potential employees for a variety of genetic traits and deciding whether to employ those found to be positive. One has to guess that not everyone being tested knows what is going on. In short, a knowledge of genetics is not a luxury, a hobby for those with the leisure and education to read. It is a necessity in a world where in the name of good medicine, people's genetic constitutions are being investigated and the information is being put to some sort of use. Perhaps the uses are mainly good, but people don't always want good to be done without knowing the meaning of what's being done. Genetic knowledge is required for informed decisions.

Let's look at the pre-employment screening in which a company declines to hire people with particular genetic qualities. There are some profound social issues lurking here. The company's interest in finding and rejecting susceptible people may simply be a means to avoid making their environment safe for everyone. If three-quarters of the population could be injured by the products of industrial or commercial activities, there is little question that regulation would be called for. But what if it were one-fourth, one-sixteenth, one in a hundred? The point is, our newly acquired knowledge of genetic variability poses new questions about the balance of obligation to individuals and to society.

A knowledge of genetics has another impact on social attitudes. Perhaps it is the most important one. Knowing some genetics can contribute in general to a tolerance for diversity and in specific to social acceptance of the disabled. Human beings instinctively shy away from great discrepancies. They show uneasiness, anxiety, even fear, in the presence of severely disabled people. But a good grasp of genetics, even a fairly rudimentary knowledge of how genes contribute to variability, makes clear that those people that everyone can agree are disabled or malformed are simply on the outer edges in the frequency distributions that include us all. There are no points in the distribution of measurements of height that divide us into dwarfs, giants, and people of normal height. There is simply a continuous gradation of heights, going from the most to the least. So, an understanding of variation, of differences between people as products of inherited differences that have been modulated by highly variable experiences, must help to induce some tolerance for diversity, some acceptance of the common

biological substrate of the human species. If we could achieve such a tolerance for the disabled, acceptance of the slight differences between ethnic groups might follow. The fact is that many of the most salient differences among us stem from varieties of experiences rather than varieties of genes.

CONCLUSION

Knowing something about genetics can teach you how the human mechanism works; how you reproduce and maintain continuity with the members of your own species, and with those of others; how you fit into the grand biological scheme; and how dependent you are on other forms of life. It can teach you that to know yourself is to make important and sensible decisions about the preservation of your individual health and that of your species. It can teach you to accept and to tolerate diversity as the rule rather than the exception. Differences are our strength, not our weakness, and it will be through the nourishment and cultivation of variety that we will survive as a species. Variety really *is* the spice of life.

FAMILIES:
Living with Cystic Fibrosis

"What is the most common, lethal, inherited disorder in the Caucasian population?" That question was asked of 124 persons on the streets and in the stores of Helena, Montana. No one gave the right answer. When told it is cystic fibrosis (CF), most said they had heard of it and many had contributed to the Cystic Fibrosis Foundation.

Only one actually knew a person with the condition. That is not surprising. Only five people with the disorder are known to live in the Montana counties of Lewis & Clark, Powell, Jefferson, and Broadway, which have a total population of 59,600.

The description of CF by the Cystic Fibrosis Foundation as "the most common, lethal, inherited disorder in the Caucasian population" raised eyebrows everywhere. But when people were told that this means an occurrence at birth

of 1 in 1800, most replied with relief, "Well, then, it can't really happen to me."

In this story, we will hear from a Helena family to whom cystic fibrosis did happen, the Laxalts of North Prospect. Theirs is a sixteen-year journey through agony and triumph, pain and joy, despair and hope. This journey finally ended with acceptance, courage, and strength in the face of continuing uncertainties.

The blue-and-white trucks of the Laxalt Construction Company are well-known in and around Helena. Bob Laxalt became company president ten years ago, when his father retired. Many residents think Bob is the most progressive member of the school board. He was recently elected to the board for a third term.

Mary and Bob have been married for eighteen years. Mary has been a nurse at St. Peter's Hospi-

tal for almost nineteen years. She met Bob there when she cared for him after a truck accident. Over the years, Mary's cheerful and gentle competence has earned her the love and respect of hundreds of citizens. Few of those people ever knew of her private anguish.

It began years ago with the long-awaited birth of their first child, John. At that time, the Laxalts were not doing well financially. They had planned on only two children. Bob greeted his son with pride and joy as his possible successor in the family business. The infant was a hefty eight-pounder. But shortly after birth, he was not passing any stool and his abdomen was swelling alarmingly. Physicians recommended immediate surgery. The doctors removed almost twelve inches of intestine during the operation because of a severe intestinal obstruction. The baby recovered

1. But it can happen to you

The number 1 in 1800 means that one baby out of every 1800 babies born alive will inherit the genetic disorder, CF. One out of 1800 is small and may seem of no consequence. But, in fact, people with CF are born and die every day.

Actually, the chance figures vary for different groups in the population. The chance of 1 in 1800 applies to all live births in the U.S. The chance for Whites is higher than that, and CF is uncommon among Blacks and Asians.

slowly from surgery. He was almost two months old before Mary and Bob could take him home.

Though John's wound had healed, he was not well. Mary worried because he continued to pass large, greasy, foul-smelling waste materials and failed to gain weight. At first, the doctors thought that too long a segment of intestine had been removed. They thought that the infant was unable to absorb all the nourishment he received. "It's something he's going to have to live with the rest of his life—it's nothing we can treat," the doctors said. Other doctors thought the baby might be allergic to his food. But many changes of formula only seemed to make matters worse.

Bob and Mary's marriage began to show the first signs of strain from John's illness. Mary's frustrations took her from one doctor to another. She seemed to be involved in the baby's care day and night. She didn't trust her infant to babysitters. When John began to have attacks of pneumonia, Mary had him sleep with her at night so she could watch him constantly.

"That was to be the last straw that drove me out of the house," said Bob, somewhat sheepishly when this reporter interviewed the Laxalt family recently.

"It was an awful time in our lives," Mary explained sadly. "During the first six months of his life, John was in the hospital five times. What savings we had were gone in no time. We couldn't keep up the payments on the house and the car, and we had to move into a trailer. The doctors couldn't tell us anything about John's condition or what was going to become of him. I was totally frustrated. I got angry at Bob for being unable to help, or seeming not to care, or not wanting to help with John's care. When he moved out, I was so lonely and angry I started to overeat, and smoke, and drink too much. I quit my job at St. Peter's. But it seemed that I was spending all my time there, anyhow, with John. I was just not getting paid for it.

"And then, one morning in February, when John was seven months old—he had just gotten out of the hospital after another bout with pneumonia and bad liver disease—Bob came home. He wrapped up the baby and said, 'I've found a doctor in Spokane who thinks he knows what's wrong with John, and I'm taking him. If you want to come along, you'd better hurry up!'"

Mary's tears began to flow as she recalled: "Bob drove like a madman, muttering under his breath all the time about not having taken John earlier. In Idaho it started to snow. When we got to Fourth of July Pass, we could hardly see. We didn't see the sign that said the pass was closed. We almost made it to the top. But then we slid off into the ditch. We were stuck for almost six hours. A snow plow finally got us out, but we had to turn back.

"We only got to Kellogg in Idaho before it got dark and started to snow again. The only motel room we were able to find wasn't heated. The baby was burning with fever, but we couldn't get a doctor to come. When the snow let up a little, we started to drive back home. Outside Missoula, John started to choke and turn blue. He died in the hospital emergency room in

2. CF can strain family relationships

Many families learn to deal effectively with CF and some even find that facing difficulties together strengthens family ties. Dealing with the psychological and social (we use the word psychosocial to describe this interaction) aspects of CF for both the family and the affected person is as important as dealing with the physical aspects. A major factor in dealing successfully with CF, or any other genetic condition, is a thorough understanding by everybody involved. At this stage in our story, Mary and Bob do not know the cause of John's physical problems.

Some families solve their problems with CF more easily than others. It is common in CF families for the mother-child relationship to become overly close and exclusive. Often the father is driven away from the care of the CF child. The mother becomes overwhelmed by the burden of care. And she may resent her husband's seeming lack of interest and involvement. The stage is set for marital conflict.

In one group of 99 families with CF children, the divorce rate was 9.5 times greater than would be expected. Another study of 214 CF families turned up a rate 5.5 times greater than anticipated. That is still significantly higher than expected, but it is 42% lower than the rate of the other sample. We cannot know for sure how much higher the rate actually is. But it is fair to say that, for CF families, the divorce rate is significantly higher than the average rate.

Missoula. When he died, he weighed only one pound more than he did at birth. We refused to have an autopsy done, and we took his little body home to be buried."

At this point Bob went on: "That really was the low point in our lives. Each of us blamed the other for John's death. But our grief finally brought us back together. What was left was a great bitterness at the doctors. They didn't make a correct diagnosis and didn't refer us to someone who could have started the right treatment.

"When we asked the doctors what would happen if we had another child, they said, 'Lightning never strikes twice.' And when Sarah was born a year later and was normal, we believed them. We thought we had put that agony behind us for good. But then. . . ." Here Bob Laxalt's voice trailed off, but Mary, with renewed strength, continued.

"Then, almost fifteen years ago, our Lisa was born. Things went well in the hospital and on the sixth day we came home. Lisa loved her bottle and had a good appetite. About two weeks after we were at home, I started to notice that her stools weren't normal. It wasn't as bad as with John, and for a long time I tried to hide the fact from Bob. But then one day I did tell our new pediatrician. She took one look at Lisa and had a stool fat test done. She told me Lisa had steatorrhea. That means too much fat in the stool because not enough fat-digesting enzymes are produced by the pancreas. The doctor noticed Lisa was not gaining weight as she should. In fact, she was losing weight. When I told the doctor about John, she put two and two together. She told me Lisa might have cystic fibrosis.

"I was so frightened that I came close to fainting. The doctor told me that the outlook was not necessarily as bad as it had been for John. Lisa could get the missing pancreatic enzymes in a pill or a capsule. That would relieve the steatorrhea and Lisa would grow. Also, her lungs could be cared for so she would be less likely to get an infection or pneumonia. The

3. Variability of CF

The difference in the severity of CF between John and Lisa illustrates an important aspect of genetic conditions. The degree of severity of some genetic conditions runs from a few, like John, who die at a very young age, to a few who may have a normal life expectancy.

The life expectancy of individuals affected with CF has improved markedly since John's death. Of all the CF patients born in 1955, 50% were expected to live to at least age five. In 1997 the median survival age for affected individuals was 31.3 years. That means that half of affected individuals died before that age

and half lived beyond that age. The reasons for that dramatic increase in life expectancy since 1955 include:

1. diagnosis early in life, before irreversible changes can take place,
2. daily pulmonary therapy to help clear the lungs of mucus,
3. use of antibiotics to manage lung infections,
4. use of special diets and enzymes to manage digestive problems,
5. better understanding of the psychological and social effects on the CF patient and family.

doctor said that children with cystic fibrosis are doing better now than they did before doctors knew much about the disorder."

Mary poured another cup of coffee for all. Then she went on: "Telling Bob was the worst part of it. I got so upset thinking he might leave us again that I did not tell him everything. I only said, 'Lisa has a mild case of the same condition John had.' But I think Bob had suspected that something was wrong for some time. When I told him that, he got so angry he knocked a big hole in the wall with his fist. He cursed so loudly the neighbors came running, thinking we were having a fight.

"But he really had changed since John died. He calmed down right away and put a bandage on his cut hand. Then we went and got the doctor out of bed and had a long talk with her. She was really good with him. Although she hadn't seen very many cases of the disease herself, she knew enough to tell him about the causes and symptoms of the disorder. Of course, all of John's symptoms were just as she described.

"When Bob came home that night, he was calm. He went over to the cradle and picked up the baby. He sat down with her in that rocking chair over there and held her in his arms a long time, trying not to cry. He kept saying over and over again, 'Don't worry, Missy Lisa'—that's his nickname for her—'we'll take good care of you.'

"The next morning the pediatrician called a doctor who was an expert on cystic fibrosis—Dr. Becker, at the university in Madison, Wisconsin. He wasn't in, but his co-worker, Dr. John Morgan, answered the phone. The two discussed Lisa's case. They concluded that she probably had cystic fibrosis, but the diagnosis should be confirmed through a test of the baby's sweat. So we went to St. Peter's Hospital here in Helena, where the doctors told us that they did not have the equipment or trained personnel to do a sweat test. They referred us to a CF center in Denver, Colorado, where we could have a reliable sweat test performed."

"Boy, I tell you, that trip was a far cry from the last one we had taken with a CF baby," Bob cut in vigorously. "My dad, who was still living then, gave us the money for the airplane and room and board at a motel near the university hospital. Within a few hours after we got there, the sweat test was done. Soon Missy was getting her pancreatic enzymes and vitamins. She also had a diet suited to her nutritional needs. In a few days, she was already gaining weight. Another thing that helped was that there was a big CF clinic there. For the first time, we met other parents of kids with CF.

"It's such a relief to know you're not alone in this world with your problems, and that others have gone the same road before you," Bob went on. "Why, we saw CF kids of all ages there, some eighteen years old, some just newborn babies. Some of them were really sick, but many were in better shape than we would have expected. The social worker and Dr. Collins, who runs that clinic, took a lot of time to tell us about CF. They gave us pamphlets and a lot of other material. I tell you, it was quite an education!

"For me, the most helpful thing they told us was that—if the lung infections are treated—most CF kids can lead a more or less normal life. They can go to school, you know, and even do some sports! Of course, there is a

4. "Sweat test"

Almost 30 years ago, doctors discovered that the salt content in the sweat from patients with CF is abnormally high.

That knowledge resulted in a diagnostic method called the sweat test. It's not always used as a screening method at birth because newborn babies usually don't sweat enough to provide a sample.

A sweat test, when properly performed and analyzed by trained personnel, is more than 98% reliable. The excretion of high amounts of salt in the sweat of a CF baby is the reason that family and friends experience a salty taste when kissing the baby. Many times, when parents are aware of this symptom, they are the first to detect the possibility of CF in their infant.

Early treatment that results from the early detection of some genetic conditions greatly improves the chances of a long life expectancy and a fairly normal lifestyle. Doctors can detect many genetic conditions earlier—such as PKU, sickle-cell disease, and CF—if the proper tests are given. Tests for other disorders are still in the research stage, and early detection is not possible at this time.

5. Cystic Fibrosis Centers

The Laxalts took Lisa to a Cystic Fibrosis Center. These centers receive partial support from the Cystic Fibrosis Foundation. As of late 1996, the centers were providing care for more than 20,000 patients in 113 locations in the United States and Puerto Rico. The main responsibilities of the centers are the care of patients, research, education for CF families, teaching, and providing resources to conduct research.

Many genetic disorders have similar foundations and care centers. Groups of people, including doctors, nurses, and families, organize volunteer groups and foundations.

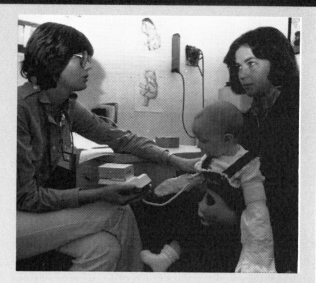

wide variability among affected people."

Mary then added, "It was really wonderful the way Bob rose to the challenge of this disease . . . or perhaps I should say condition."

"While I took a little vacation," Mary continued, "Bob read and read and read. He kept bugging the doctors and nurses to explain things to him. Why, he even started to teach me at night in the motel and to tell me about all of the research that was going on.

"While Lisa was getting good care, we spent a few days seeing the sights of Denver and the surrounding Rocky Mountain area," she added. "It really was our first vacation since our wedding."

"Well, a lot of water has gone over the dam since then," Bob said. "If we had known then all that we know now, perhaps we would not have come home from Denver in such an optimistic mood. One thing we did agree on, though, was that we were not going to have another child. We found out in Denver that CF is an inherited condition, and there is a twenty-five percent chance of CF in any child we might have. Sarah and Lisa have kept us plenty busy. We really couldn't have handled another child.

"At that time, CF couldn't be diagnosed before birth. So each time you conceived, you wouldn't know until the baby was born whether it was affected. And even though a twenty-five percent chance meant that the odds were in our favor three-to-one, Mary and I both felt a one-fourth prob-

6. Disease vs. condition

Why did Lisa's mother say disease and then say condition? Psychologically, the patient and the family are better able to cope with CF when they think about it as a condition caused by an inborn (genetic) error of metabolism. It was not something caught or caused by bacteria, viruses, or other microscopic organisms. It is not contagious. If the symptoms caused by the genetic error are treated, the patient can live a fairly normal life.

With infectious diseases, doctors often can treat the cause, not just the symptoms. They can give a patient medicine that will destroy the organism that causes a disease. But genetic conditions cannot be treated in the same way. Genetic conditions can be treated, but not cured.

The use of the word condition, rather than disease, to describe genetic abnormalities is not universally accepted. Because CF and other genetic problems demand an ongoing regimen of medical care, the use of the term disease is appropriate in the opinion of many in the health-care field. However, there is evidence that dealing with the problem as a condition, rather than a disease, has a higher success rate among some families and affected individuals.

7. Probability

How did the Laxalts know that if they were to have another child its chance of having CF would be 25%? To understand the answer, you need some background information on a branch of mathematics called probability. Much of what happens in genetics follows certain laws of probability. These laws allow us to make predictions about the occurrence of CF in the Laxalt family. The basic question in probability is,

How often should we expect a particular event to occur in a given number of events? You can answer that question if you learn how to use the following formula:

$$\text{Probability} = \frac{\text{number of events of choice}}{\text{number of possible events}}$$

Now, follow the directions of your teacher.

8. Special treatment and concerns for CF patients

All CF patients receive special treatment based on their individual symptoms. Each treatment has something to do with the extremely thick mucus that is the central characteristic of CF. We don't know the exact cause of the thick mucus, but it is probably related to the abnormal salt concentrations within the cells that produce the mucus. Treatment generally includes:

Bronchial drainage or postural drainage. This is a form of physical therapy that helps to loosen mucus from the lungs. It keeps lung passages open. The method Lisa uses may be accompanied by a chest-clapping and vibration procedure done by a physical therapist or by a family member who has learned the procedure. The procedure loosens the mucus and the patient tries to cough up as much of it as possible.

Antibiotic therapy. Some CF patients take drugs regularly to fight respiratory infections. Some take them only when there is an indication of a lung infection. We don't know why CF patients are more prone to lung infections. One possibility is that the mucus accumulation in the lungs may act as a breeding ground for bacterial growth. In addition, plugged respiratory tubules are unable to pass fluid—normally secreted by lung cells—up and out of the lung. This fluid is an excellent culture medium for bacteria. The prime bacterial infection is caused by <u>Pseudomonas aeruginosa.</u> That infection can be difficult to treat. CF patients often require hospitalization. Sometimes they receive intravenous antibiotic treatment.

Dietary management. In CF patients, thick, sticky mucus blocks the ducts that carry enzymes from the pancreas and other digestive glands. One result is the swelling and eventual bursting of the enzyme-producing glands in the pancreas, especially. The organ becomes cystic (full of enlarged glands), and when the cysts rupture, the digestive enzymes escape into the surrounding tissue and destroy it to varying degrees. The damaged tissue eventually heals and becomes scarred or fibrosed. Thus, the old name for CF, cystic fibrosis of the pancreas, makes sense. With plugged ducts and damaged tissue, the enzymes cannot get to the intestine and, as a result, much of the food is not digested and is excreted. Lost along with the undigested fats are large amounts of the fat-soluble vitamins A, D, E, and K. Thus, doctors prescribe vitamins, pancreatic enzyme replacements, and a variety of nutritional supplements. CF patients often must take a large number of pills each day. According to 1995 data (the latest available) from the Cystic Fibrosis Foundation, the *annual* cost of care for the average person who had CF was $39,166.

Infertility. Males with CF are usually infertile because of the total absence of sperm. The vas deferens, the tube system that leads from the testicles to the urethra of the penis, becomes blocked before birth, presumably due to the same sort of situation that exists in the digestive duct systems (see Dietary Management). The absence of sperm does not alter sexual function. We can't explain the decreased fertility seen in females with CF, but it may be related to the thicker-than-normal mucus in the fallopian tubes, which lead from the ovaries to the uterus; that could impede the passage of an egg, of sperm that enter the tube, or of a fertilized egg.

Lisa talks of her condition and treatment as ordinary, everyday matters. Apparently, through her knowledge and understanding, she has learned to live with CF in a positive way. But we must remember that each time Lisa is forced into the hospital could be the last. Death from a severe infection, or pulmonary or heart failure is always possible. It will happen to Lisa. She will die from her condition at a relatively young age. People affected by many kinds of genetic conditions must face the possibility of early death every day. Many, like Lisa, seem to adjust, but there are others who do not adjust well. They require special counseling and support to cope with their problems.

ability was too high a risk to take. Two kids with CF were enough!"

To this Mary added: "Now scientific advances have made prenatal testing for CF possible. I'm not sure if we would do things any differently now ourselves, but I'm glad that at least now other families have more choices."

The afternoon sun was declining. For a moment, no one spoke. Sleeper, the family's dog, dozed before the fireplace.

Almost immediately, the clock chimed. It was three o'clock. Sleeper bounded toward the front door, wagging his tail. Together, Bob and Mary said, "Time for the kids to come home from school."

At that instant, the girls burst in, leaving a trail of mittens, scarfs, jackets, books, boots, and snow on their way to the refrigerator. Lisa stuck her head around the corner. "Oh, it's Alex—you're the reporter, aren't you? Come, join us for a glass of milk and some cookies."

I joined Sleeper and the girls in the kitchen for their after-school snack. And, although this reporter doesn't like milk, we had a delightful time. Lisa seemed to be breathing a little faster than Sarah, but that may have been my imagination. The only unusual thing that caught my attention was how often Lisa coughed.

"It feels good to be home," Lisa explained, "because our house is humidified. I can breathe and sleep easier. I take enzyme capsules before supper," she went on, "and vitamins with my supper. I'm still taking an antibiotic since my last chest infection, and every morning and night I do postural drainage." She explained that this involved positioning her body in various ways to permit the force of gravity to assist in draining the mucus from her bronchial tubes.

Having to take her enzyme capsules before lunch at school is the main thing that makes her different from her schoolmates.

Lisa is in the tenth grade and takes a full load of courses. She also swims, runs, is the timer of the girls' basketball games, and plays drums in the band. Three days each week, she works as a volunteer at Shodair Children's Hospital after school.

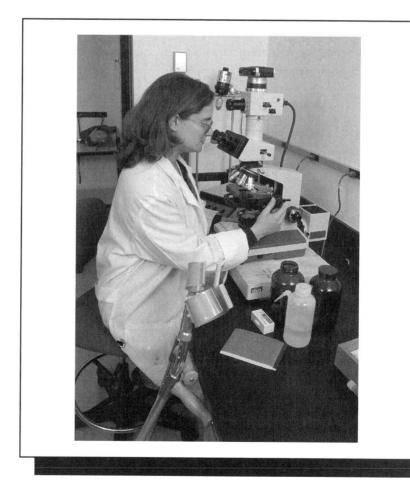

"I might as well," she quipped, "since I am one of their favorite consumers." But she went on more seriously. "No, I really am not. In the last ten years, I've been hospitalized only five times, the last time three weeks ago. When my chest gets too full, I have to do additional pulmonary therapy or have help from a physical therapist. And periodically the technicians in the lab do throat and sputum cultures to keep track of the bacteria in my system. The physicians want to know whether the bacteria are becoming resistant to the antibiotic I am taking. When I get a fever and my cough gets worse, the physicians request X-rays of my chest.

"I know pretty nearly everyone at Shodair, and they all have been very nice to me. Every now and then, I have a soda with the new geneticist there. She has been teaching me all about the genetics of CF, since we don't get much human genetics in our high school biology class.

Lisa was happy to share her knowledge with me. "CF is recessively inherited. That means that carriers, like my mom and dad, who have only one CF gene, are perfectly normal. Most carriers never have affected children because they don't marry another carrier. Since a *pair* of CF genes (she called them CF alleles) are required for a child to have CF, *both* parents must be carriers. And since both parents also have the non-CF form of the gene (allele), their normal children each have a two-in-three chance of being carriers.

I gave up at that point and told Lisa she'd lost me.

10. The discovery of genes

And maybe Lisa's explanation has lost you, too. But the following information should help you understand the basic genetics involved in CF.

Our story starts with Gregor Mendel, an Austrian monk, who among other things tended a garden at the monastery. While working in his garden, he noticed how different some of the pea plants were from one another. He had planted what seemed to be the same kind of seeds. But the plants had different stem lengths; some were short and others were long. The peas they produced were different in color and texture. Some plants had flowers at the ends of their stems (terminal flowers), while others had them in the angles formed by the stems (axial flowers). Those variations puzzled Mendel. He set out to find what caused them.

What Mendel discovered and explained to the world in 1865 became the foundation for the science of genetics. All living processes are controlled by factors (we now call them genes) that express themselves during the development and life of the individual. These genes are inherited from parents and transmitted to offspring. Human genes distinguish humans from nonhumans. Genes also characterize certain groups of humans, certain families, or certain individuals—like Lisa with her CF condition.

Mendel spent several years raising plants that always produced offspring that showed the same form of a trait. He produced pure varieties of pea plants. For example, one variety always produced wrinkled seeds. Another variety always produced round seeds.

Mendel then began experiments involving crosses of the pure varieties of pea plants. He made hundreds of crosses to study a particular trait. One was a cross between pea plants whose seeds were always round and plants whose seeds were always wrinkled. When a cross is made between two varieties, the parent generation is called the P_1 generation. The offspring of the P_1 cross are the first filial, or F_1, generation.

In every case, Mendel found that all the F_1 plants had round seeds. When only one form of a trait (in this case, round seeds) appeared in the F_1 generation of a cross of pure varieties, Mendel called that the dominant trait. The trait that did not show was called the recessive trait. The offspring in all of Mendel's first crosses were exactly like one of the parents. There were never any in-between types nor any that showed a mixture of the parents' traits.

Mendel then crossed many of the F_1 plants by allowing self-pollination to occur. In the next generation, the F_2, the dominant trait (round seeds) appeared in 75% of the offspring. In 25%, the recessive trait (wrinkled seeds) reappeared. That gives a ratio of three dominant traits to one recessive trait in the F_2 generation. Mendel observed the same results in crosses involving other traits.

Mendel looked for a hypothesis to explain those results. He assumed that some part of the reproductive cell carried something that transmitted traits to offspring. We now call those parts genes, and we know they are carried on chromosomes. Mendel assigned letters to stand for the genes that caused each trait. For example, he used a capital R to stand for the gene that caused the dominant trait—round seeds. A lowercase r stood for the gene that caused the recessive trait—wrinkled seeds.

In this example, we are assuming that the appearance of round seeds is caused by a certain gene, R. To characterize the trait itself, we use the word phenotype. We use the word genotype to refer to the combination of genes that is related to the pheno-

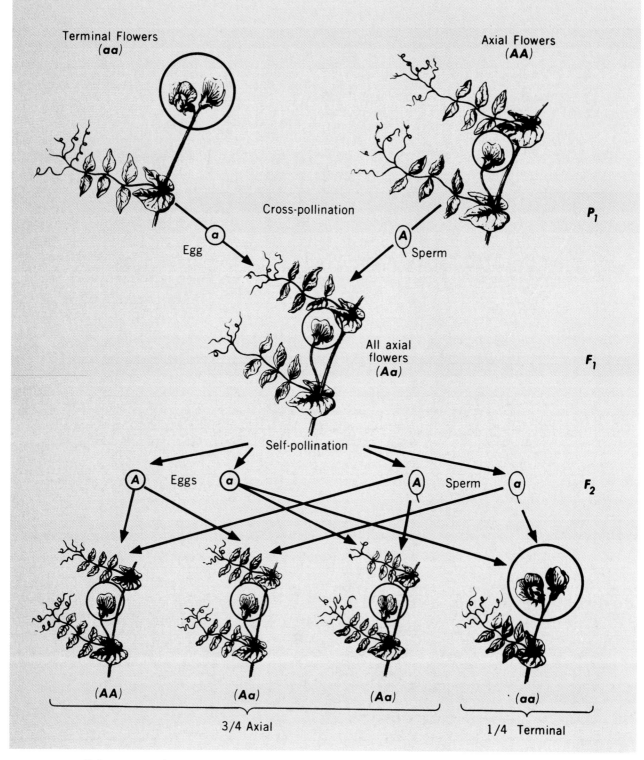

FIGURE 1 ■ Inheritance of terminal and axial flowers in peas.

type. The egg and sperm that combine to produce the offspring may be genetically different. Each male or female reproductive cell contains only one of each kind of gene. When the two cells combine, the new cell contains two of each kind of gene—one from the mother and one from the father.

Mendel suggested that the reproductive cells of a pure, round-seed plant would carry only R genes. Thus, its genotype would be RR. And those of a pure, wrinkled-seed plant would carry only the r genes, genotype rr. We describe these individuals as homozygous, because both alleles are the same, RR or rr.

The result of a cross between pure, round-seed plants and pure, wrinkled-seed plants would be an F_1 generation of plants that had both R and r genes. The genotype of those plants would be Rr. When the alleles are different (Rr), the individual is said to be heterozygous.

Because the R gene exerts a dominant effect, the phenotype of the F_1 generation would be round seeds. But we know that the r gene was present in the F_1 plants, because the wrinkled-seed trait reappeared in the F_2 generation. Figure 1 shows how this works for another set of characteristics in peas.

Gathering knowledge about, and developing an understanding of, the genetics of nonhuman life has been going on since before recorded history. Understanding human genetics is in some ways more difficult. In the laboratory, geneticists prefer to work with pure lines of organisms whose offspring will always show the same form of a trait. They obtain those

pure lines by breeding plants and animals that are closely related for many generations. They repeat crosses many times and try to obtain many offspring. They keep the experimental organisms in an environment in which heat, light, nutrition, and other factors can be controlled carefully.

Of course, it is not possible to carry out such experiments on people. But geneticists have learned a great deal about human genetics in other ways. One way is by tracing the inheritance of traits. They observe the distribution of a trait in as many generations of a family as possible. Then, they put the information in a diagram called a pedigree. Figure 2 is a pedigree for a family in which CF is present.

Study the pedigree of the Laxalt family in Figure 2. The grandparents are not affected, but Mary and Bob, the parents in the next generation, are both carriers. At that time, there was no test to detect a carrier of CF. The fact that the parents in this pedigree are carriers could not become apparent until they had their first affected child, John. The second child, Sarah, did not have the disorder. We now can determine whether Sarah is a carrier. If she is a carrier, CF could occur in her children only if the father also is a carrier. We now can analyze DNA obtained by a blood test to determine in most cases who in a family are carriers for CF.

We can reason that either Michael or Emma must have been a carrier. Either Herman or Ann also must have been a carrier. How do we know that?

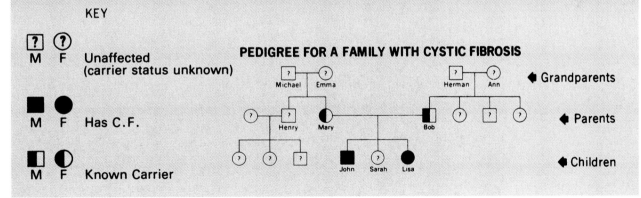

Figure 2 ▪ Pedigree for a family in which cystic fibrosis is present.

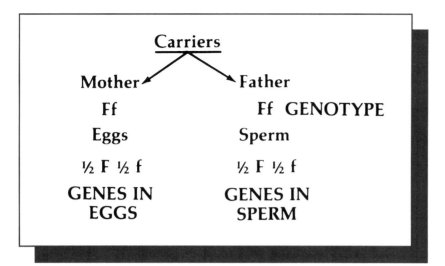

Carriers

Mother → ← Father

Ff   Ff **GENOTYPE**

Eggs   Sperm

½ F ½ f   ½ F ½ f

GENES IN   **GENES IN**
EGGS   **SPERM**

FIGURE 3

"Maybe it'll help if I draw you a picture," Lisa said. Figure 3 is what she drew.

I had to interrupt her again to find out why both members of the pair of genes did not end up in the same egg or sperm.

"Oh, no, they never do," she said. "The members of the gene pair are on different chromosomes. Each member is on one of two paired chromosomes. An egg or sperm has only twenty-three chromosomes—one of each of twenty-three pairs. That way, when the egg and sperm meet, the new cell will have the correct number of chromosomes—forty-six—and the twenty-three pairs match. If the gene pairs were not separated in the egg and sperm, we would double the number of genes in every cell of our bodies in each generation." That impressed me as a reasonable argument. But it did not explain how these two genes got sorted out from each other to end up in two different cells. So, Lisa began to tell me about meiosis.

Lisa quickly drew a picture of a person (Figure 4), and labeled it Body. Then, she drew a picture of a heart (Figure 5), next to which she wrote Organs.

"The organs all consist of cells," she said. And she drew several kinds (Figure 6).

All cells, except the red blood cells, have a nucleus. In the nucleus of a large cell, she drew many wiggly lines. She called these strands chromatin (Figure 7).

The chromatin in a nondividing cell is all of the material of the chromosomes of the cell in a sort of unraveled state. That makes the chromosomes difficult to see except during cell division. From her social studies book, Lisa pulled out a beautiful photo that the new geneticist at Shodair had given her. It showed her own chromosomes from a dividing cell (Figure 8).

"Is that a karyotype?" I asked her, using the only technical term in biology that I thought I understood.

"Of course, it isn't," she said. "People don't have karyotypes. They have chromosomes. The chromosomes are given numbers according to their size and the pattern of light and dark bands that

each of them has. Chromosomes contain the genes—lined up inside them, or on them, sort of like beads on a necklace. Many thousands of genes are packaged into only twenty-three pairs of chromosomes. And one member of each pair comes from the mother; the other comes from the father. It's only when you cut out the chromosomes in this picture and line them up in this order that you make a karyotype." She carefully removed a karyotype from between the pages of her Spanish book (Figure 9).

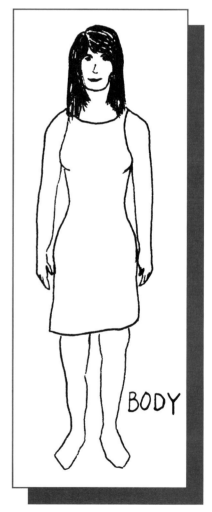

FIGURE 4 ▪ Lisa's drawing of a person.

18

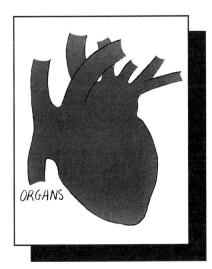

FIGURE 5 ▪ Lisa's drawing of a heart.

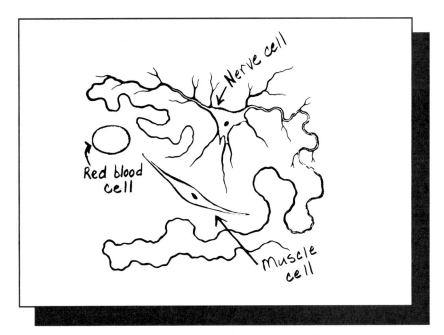

FIGURE 6 ▪ Lisa's drawing of different kinds of cells.

As I studied the karyotype, she continued, "Now, you must imagine that the number seven chromosomes contain a pair of CF genes—one gene on each of the two chromosomes. When most body cells divide, the chromosomes are first doubled in number and then the nucleus divides. Each of the two new nuclei has the same number of chromosomes as the first cell did. That process is called mitosis. Through mitosis, the number and kind of chromosomes in most kinds of body cells remain the same. So, each of my body cells has a pair of CF genes in it.

"But the process is different when reproductive cells—egg cells and sperm cells—are formed. The process of chromosome duplication and nuclear division is called meiosis. It differs from mitosis in several ways. One of the differences is that the cells end up with half as many chromosomes as the original cell had (instead of the same number as with mitosis). Producing the sperm or the egg cell requires two divisions. Let me draw you a diagram. To make the drawing simple, I'll just show one pair of chromosomes instead of all twenty-three pairs. And I will mark a place on each chromosome to stand for the CF gene (Figure 10).

"My mom or dad's cells in meiosis would be diagramed differently. Because they are carriers, only one of the two chromosome 7s would have the CF gene." Lisa proceeded to sketch the situation in her mother and father (Figure 11).

"In the first cell division, the chromosomes line up in pairs. Then, they double and separate. When the cell divides,

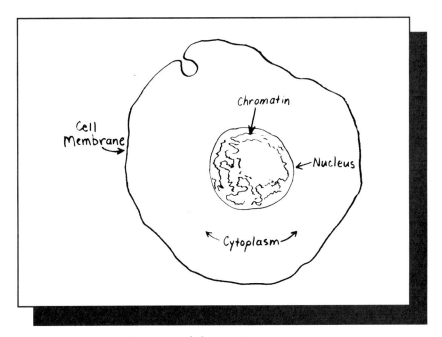

FIGURE 7 ▪ Lisa's drawing of chromatin.

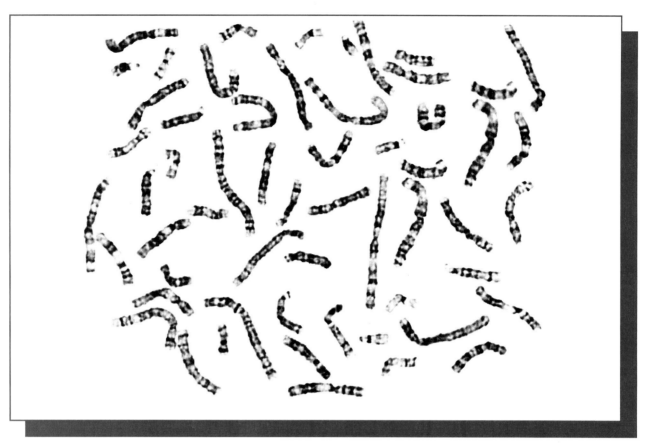

FIGURE 8 ▪ Human chromosomes. The banding patterns result from the use of a special stain. (Courtesy of Molecular Diagnostics, The Hospital for Sick Children, Toronto, Canada.)

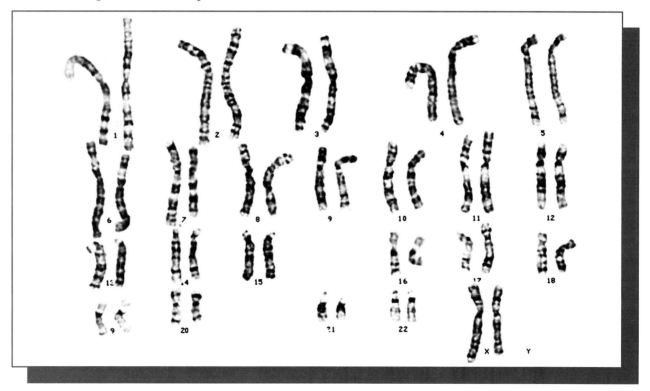

FIGURE 9 ▪ Normal human female karyotype. (Courtesy of Molecular Diagnostics, The Hospital for Sick Children, Toronto, Canada.)

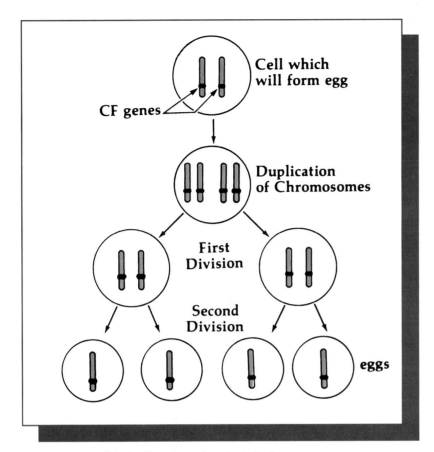

FIGURE 10 ▪ Lisa's diagram of meiosis in her eggs.

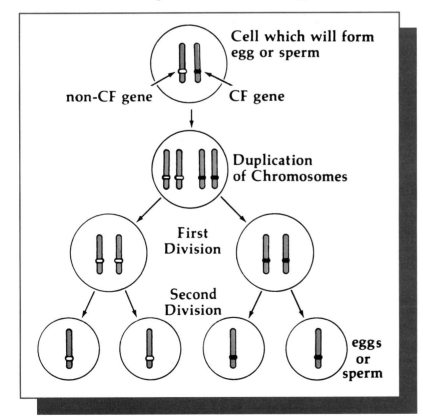

FIGURE 11 ▪ Lisa's diagram of meiosis in her mother and father.

each new cell contains twenty-three pairs of chromosomes. These new cells then divide, and the chromosome pairs are separated. That's right, only twenty-three chromosomes, not forty-six. That's why each egg and each sperm can contain only one gene that may or may not result in CF."

The force of Lisa's argument was convincing. But it took a little more study for me to take it all in and to learn all the new words.

While Lisa was explaining meiosis to me, Sleeper kept nudging me for my leftover chocolate chip cookie, which he swallowed in one gulp without chewing. He also noticed my untouched milk glass, and I quietly emptied it into his dish.

Now I have a new friend. Whenever we meet in town, he jumps on me with much tail wagging and playful barking. I have never seen his eyes, but I'm told they are a lovely golden brown. As I scratched Sleeper's ears, I said to Lisa, "So, CF is caused by a recessive gene and not by a dominant gene. Right?"

Lisa responded thoughtfully to my question. "Well, Alex, yes and no. It depends on how you look at it. If you're using the word dominant the way we do in ordinary conversation, then you are wrong. But if you mean dominant as the characteristic that is expressed even if only one allele of that type is present, then you're right."

By this time it was dark outside, and Mary had dinner ready. Bob returned and, looking over our studies, kissed Lisa on the head: "At it again, Missy? That's good; the best way to live with a condition is to understand it." I gladly accepted an invitation for

11. Meiosis

If we look at Lisa's two diagrams of meiosis, maybe her explanation will be a little easier to understand. Of course, the cell that will form her eggs has 23 pairs of chromosomes, but she is diagraming only the pair for chromosome 7, where we find the genes for CF. The chromosomes duplicate and then separate. Each of the two new cells contains one pair of chromosomes. These new cells divide, and the chromosome pairs are separated. So, the number has been reduced by one-half. In Lisa's diagram—from two to one, in her cells from 46 to 23.

The same process goes on in her mother's egg-producing cells and her father's sperm-producing cells. The difference is that all of Lisa's eggs contain a chromosome with the CF gene. But only one-half of her mother's eggs and her father's sperm contain the CF gene. The other half do not.

dinner. Mary had prepared delicious broiled fish and fresh fruit.

During dinner I asked Bob the question he must have been asked hundreds of times. "No, we are not related to Paul Laxalt, the former U.S. senator from Nevada, although both families are of Basque descent."

He explained that the ancestors of both families had come to the West as sheepherders. Bob's grandfather had been an independent sheepherder near Dillon, Montana. But after the First World War, he sold his ranch and began the construction firm in Helena. When I asked if CF had anything to do with Basque ancestry, he grew thoughtful.

"Maybe, but I'm not sure. I know several other Basque families who've had children with CF. The best known of course, is the late State Senator Etchart of Glasgow. Etchart had been very active in CF volunteer work and was responsible for the passage of a law in Montana that provides direct financial support to families being drained by high medical expenses. The research and library assistant at Shodair did a search of medical literature for me on a possible relationship between Basque and CF, but we came up with nothing. However, that's something I'm going to do some research on—if I ever can get away from the office long enough." The last remark drew several laughs and good-natured remarks from around the table.

Over dessert Lisa took me on again. "I'm not going to let you get away until you understand why the unaffected brother or sister of a person with CF has a two-

12. Dominant and recessive characteristics

Lisa's point about dominant and recessive genes is a subtle, but important distinction. We talk about dominant and recessive genes only as a matter of convenience. Actually, it is the characteristic (phenotype) that is dominant or recessive—that is, expressed or not expressed. Saying dominant gene is a quick way of talking about a gene whose effect is expressed (the phenotype) even if only one of that kind of allele is present. Recessive gene is a quick way of referring to a gene whose effect is not expressed in the phenotype unless *both* alleles are of the same type.

In everyday language, we use the word dominant to mean controlling or overpowering something else. But genes do not work that way. Genes provide the instructions for making certain cellular products—usually proteins. A gene "codes" for a particular protein. If a cell contains a so-called dominant gene, the protein that will be evident in the phenotype is the one coded for by that gene. If a recessive gene is present along with a dominant gene, the recessive gene will produce a gene product that is different in structure and function from the product of the dominant gene. But the product of the recessive gene will not be evident in the phenotype. If two recessive genes are present, the only protein produced will be the one coded for by those genes, and that protein will be expressed in the phenotype. Therefore, the terms dominant and recessive do not refer to one gene dominating another. Rather, they refer to the allele or alleles whose product is expressed in the phenotype.

Lisa should also remind Alex that, while CF is inherited as a recessive characteristic, some genetic disorders are inherited as dominant characteristics.

13. Genetic disorders in ethnic groups

Some recessive disorders occur more often among certain populations or ethnic groups. Tay-Sachs disease, for example, is more frequent among Jewish populations that trace their ancestry to northeastern Europe. We assume that the gene for Tay-Sachs arose by mutation in that population many generations ago.

Religious and cultural traditions have resulted in a tendency for individuals from that population to marry within the group. That is not unusual in human populations. The result is an increased probability that individuals who are heterozygous for the Tay-Sachs gene will marry. That increases the probability that they will bear a child with Tay-Sachs disease.

Some populations are so restrictive in their matings that the frequency of certain disorders is extremely high. These populations, called isolates, have few marriages with individuals from outside the group. As a result, many marriages take place between closely related individuals—inbreeding. Closely related persons are much more likely to have the same genes—harmful or not—than unrelated people. When those genes combine in their offspring, the likelihood of a genetic disorder is higher. The Amish, for example, have a high frequency of a certain type of recessive dwarfism.

The increased probability of genetic disorders that result from inbreeding probably is one of the reasons for the long-standing cultural taboos against incest. The Bible (Old Testament) cautions that "none of you shall approach to any that is near of kin to him, to uncover their nakedness" (Leviticus 18:6).

There does not seem to be any particular ethnic group in which the frequency of CF is higher than in other groups. However, CF occurs most often among Whites, seldom among Blacks and Asians. As the following describes, we now know that 70% of CF mutations are all of the same type of change in the CF gene. (See box 19 for more information about CF mutations.)

thirds chance of being a carrier." I groaned a bit, but submitted humbly to my young teacher, who drew another picture (Figure 12).

Then, she explained the picture. "You can see that half of the eggs and sperm carry the CF gene (f) and half carry the normal gene (F). This scheme is called a Punnet square. It gives you all the outcomes possible at each conception. There are only three: ff—the affected; Ff—the carriers; and FF—the normal, noncarrier persons. And they occur in a ratio of 1/4:1/2:1/4, right?"

"Right," I responded. "So, how come the two-thirds chance?" I thought about it for a moment, and it became clear. One-fourth are affected—they have the CF phenotype. Three-fourths are not affected. Theoretically, one of the unaffected three would have the FF genotype—no recessive CF gene. The other two would have the f gene in their genotypes.

Thus, we could predict that two out of three without CF are carriers—a two-thirds chance.

I heaved a great sigh of relief. I had just learned some basic ideas of genetics in a most agreeable and meaningful manner. I thanked Lisa and rose to leave, but she was relentless.

"One last thing, Alex. You should learn a little bit about the Hardy-Weinberg principle before

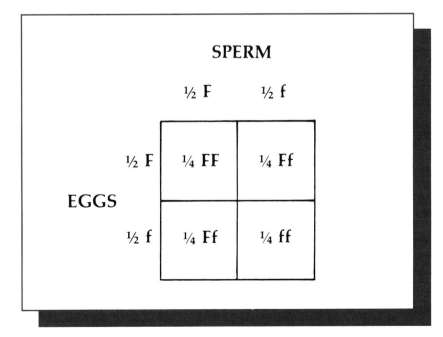

FIGURE 12 ▪ Punnett square.

14. What are the chances?

The 1/4:1/2:1/4 ratio of genotypes is only what we might *expect for each pregnancy* when both parents are CF carriers. Let us suppose that these parents plan to have four children. For each child the chances are the same: one-fourth that it will be normal, one-half that it will be a carrier, and one-fourth that it will be affected with CF. It does not mean that we would expect one of the four children to have no gene for the CF condition, two to be carriers, and one to have CF.

Look back at the pedigree in Figure 2 on page 17. The carrier parents in that pedigree had three children. Two of them had CF. There is a case of a family with eight children, five of whom have CF. Or, there could be a family of eight children, none of whom would have CF, even though both parents were carriers. Probability tells us only what we might expect *each time* a child is conceived by parents who are CF carriers. Each child has the same one-fourth chance of having CF.

you can really understand genetics." Now I was really excited, for I had not expected anything more. Mary poured me another cup of coffee, and I was ready for this last challenge.

"I don't think it is necessary to go into all the detailed mathematics of the Hardy-Weinberg principle," Lisa said gravely, and without the slightest hint of condescension, "but an understanding of what we call 'gene frequencies' in a population is important. Let's use the figure 1/1600 for the frequency of CF among White people. I should use 1/1800, the frequency for all live births in the U.S.," she noted, "but the arithmetic is harder. Now, if we have a population of 1600 newborn White children, we can predict the frequency of the genotype in this population. Tell me, what is the frequency of genotype ff?"

I thought for a moment and said, "1/1600."

Lisa laughed and said, "You are a fast learner, Alex," and wrote: ff = 1/1600. Then she continued.

"The gene pool is the term geneticists use to describe all the genes that the members of a population may contribute to the next generation. There are just two al-

15. Clearing up a misconception

Lisa knew that there is a common misconception about recessive traits. She wanted Alex to understand the Hardy-Weinberg principle so he wouldn't get the same idea. Some people assume that, because a trait is recessive, the gene eventually will disappear from the population.

The fact is that recessive genes will not disappear unless some forces are operating that change the frequency of the recessive gene. For example, pea plants survive and reproduce perfectly well whether they have round seeds or wrinkled seeds. Therefore, the frequency of the recessive gene does not change over time, assuming certain, special conditions, such as the absence of mutation.

In 1908, two scientists, G.H. Hardy, an English mathematician, and W. Weinberg, a German physician, worked out this principle. They showed that the frequencies of genes and the genotype ratios remain constant within a population. That principle holds true in the absence of forces, such as nonran-

dom mating or natural selection, that can alter the frequencies. Hardy, in his 1908 paper, stated that "there is not the slightest foundation for the idea that a dominant character should show a tendency to spread over a whole population, or that a recessive should tend to die out."

Of course, with CF, there are forces at work that change gene frequencies—there is selection against the CF gene. Not long ago, few children who were born with CF lived past infancy. Even now, many people with CF decide not to marry, and many of those who do, choose not to have children. Also, about 98% of males with CF are sterile, according to 1997 data from the Cystic Fibrosis Foundation. But, there are many more heterozygotes than homozygotes. For that reason, it would take perhaps hundreds of generations to reduce the frequency of the gene very much. So, we can expect the frequency of the gene that causes CF to diminish, but not very much.

leles for CF—the F allele and the f allele. In the gene pool, a certain proportion of the alleles are F and the rest are f. The proportions must add up to one hundred percent, or one. In this case, one means all or the whole.

"Using the formula developed by Hardy and Weinberg and the one piece of data we have, ff = 1/1600, we can calculate the frequencies of the genes F and f in the gene pool. Once we know that, it's easy to calculate the frequencies of the other genotypes, Ff and FF, in the population. The key step is to determine the frequency of the recessive gene, f, in our population. We do that by taking the square root of the frequency of homozygous recessives. The square root of 1/1600 is 1/40. This is the frequency of the recessive gene: f = 1/40. Earlier I said that the frequencies of the recessive gene (f) and of the dominant gene (F) must add up to one hundred percent, or one. So, knowing that, how can I get the frequency of the dominant F gene?"

I am not great when it comes to math, but it seemed almost too simple to be true. I said, "If the frequency of the recessive gene plus the frequency of the dominant gene must equal one, and we know the frequency of f is 1/40, then the frequency of F must equal 1 minus 1/40—which is 39/40." Again, Lisa laughed with glee, because her student seemed to be learning very rapidly.

Lisa continued, "Now, in each population there are two gene pools that must be combined into one. Each population has a male gene pool (in the sperm) and a female gene pool (in the eggs). The frequency of the f gene in the sperm and the eggs is 1/40, while the frequency of the F gene is 39/40 in the sperm and the eggs."

"One of the laws of probability states that 'the chance of two independent events occurring at the same time is the product of their chances of occurring separately.' When we combine the gene frequencies of both the male and the female gene pools, we get the proportions of FF, Ff, and ff individuals among all of the offspring.

"As we determined earlier, the frequency of the F gene in both the male and female populations is 39/40. The FF genotype is the result of chance union of an egg and sperm each carrying an F gene. Using the second law of probability, this is 39/40 x 39/40 = 1521/1600. So, 1521 out of our population of 1600 will have the FF genotype.

"The Ff genotype can occur in two ways: F sperm can combine with f eggs, or F eggs can combine with f sperm. To calculate the Ff genotype, therefore, we must take

16. The Hardy-Weinberg principle

To understand the Hardy-Weinberg principle more thoroughly, let's look at another trait and determine its gene and carrier frequencies. The ability to taste a chemical called PTC is inherited. To some people the substance tastes bitter. Others cannot taste it.

In one class in an American school, the PTC test showed 30 tasters of PTC and 10 nontasters. Each student took home a small amount of PTC solution and tested her or his parents. Their results appear in Table 1. From these data, can you determine which of the traits is dominant and which is recessive?

When parents who have the dominant phenotypes are crossed, the recessive trait can appear in their offspring. But when both parents have the recessive phenotype, all their children also should have the recessive phenotype. In this test, only the nontaster by nontaster matings gave that sort of result. Therefore, the nontaster trait must be recessive. Tasters can be either homozygous dominant (TT) or heterozygous (Tt). Nontasters must be homozygous recessive (tt).

Now, let's consider a population in which 64% of the population are tasters and 36% are nontasters. What are the frequencies of the gene? What are the frequencies of the genotypes?

What is the only genotype frequency we already know? With that value, what gene frequency can we determine? Because we know that the frequency of T and the frequency of t must equal one, what is the frequency of the dominant gene? Now, calculate the genotype frequencies of the carrier and the homozygous dominant.

Parents	Student Tasters	Student Nontasters
Taster × taster	20	2
Taster × nontaster	10	5
Nontaster × nontaster	0	3

TABLE 1 Results of PTC test.

2(39/40 x 1/40). We multiply by two because the Ff genotype can occur in two ways. That works out to 78/1600, or about 1 out of 20. That means that almost five percent of the White population are carriers of CF," Lisa concluded.

Lisa's calculations surprised me, and I redid the arithmetic several times to see how she had reached five percent from 1/1600. I finally had to concede that she was right. Then, Lisa went on to tell me that there is a quick and easy way to calculate the carrier frequency for most recessive disorders: "Since the gene frequency (f) for most is small, 1 in 40 for the CF gene, F will be so large that one can consider it to be approximately one. Now, the frequency of the Ff genotype, the carrier frequency for CF, becomes 2(~1 x 1/40) or ~2f. In other words, the carrier frequency is about double the gene frequency. As long as the incidence of the disorder (f^2) has a fairly easily calculable square root, you can do the carrier frequency in your head.

"Try it with another example. We learned in school about a recessive disorder called PKU, a form of inherited mental retardation. It appears in about 1 in 10,000 live births. The carrier frequency? Why, it couldn't be easier! The square root of 1/10,000 is 1/100 and 2 x 1/100, obviously, is 1/50, and that indeed is a very close estimate of the actual carrier frequency for PKU. Isn't it interesting that the carrier frequency seems so high (1/20 for CF; 1/50 for PKU) even when the incidence of the disorder is relatively small? Now, look at the picture of this iceberg," Lisa urged me as she drew another picture.

Figure 13 represents all the f genes in the population. As you can see, for each affected person with two f genes, there are seventy-eight unaffected carriers with a single f gene. The rarer the recessive condition, the greater is this ratio of carriers to affected persons. In a rare condition occurring once in a million, there are almost two thousand carriers for each affected person.

At this point, I asked a question that had been in the back of my mind. I asked Lisa, "Can we figure out what the chance would be for two unrelated carriers, like your mom and dad, to meet and marry?"

"That's easy, Alex. We know

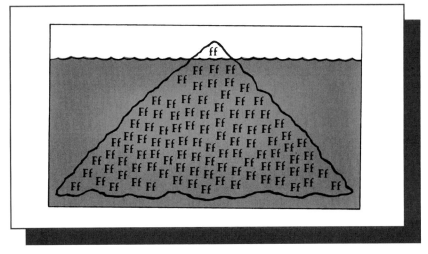

FIGURE 13 ■ Lisa's drawing of the f genes in a population.

17. Carriers of CF

It was easy for Lisa to do the mathematics to answer Alex's question. As her father had said earlier, "The best way to live with a condition is to understand it." Lisa can cope with her condition because she has worked hard to understand it. But how about you? Do you understand it? Let's review what Lisa just explained.

Bob and Mary Laxalt are carriers of CF. Their genotype, Ff, will be found in 1 out of every 20 people in their population. What are the chances that two people who are carriers will marry? The probability that two independent events will occur together is the product of their individual chances of occurring separately. Now, with the background you have in probability, answer the following questions.

1. What is the probability that a woman is a carrier for CF?
2. What is the probability that a man is a carrier for CF?
3. What is the probability that two carriers will marry?
4. What is the probability that two carriers will marry and have a child with CF?

18. An increased probability

Something seems wrong here. We just looked at the probability that two White people are carriers and at the probability that two such people will have an affected child. We found the probability to be 1/20 x 1/20 x 1/4 = 1/1600. What is different about the question we're now asking?

What is the chance that a sibling of an affected person will have an affected child? The math given is 2/3 x 1/20 x 1/4. Where does that two-thirds come from? Look again at Figure 12 on page 23.

Suppose you are the brother or sister of someone with CF, like Sarah in our story. You don't have CF, so your genotype is not ff. But it could be FF or Ff. Of the remaining three chances in four, you have one chance of having FF and two chances of having Ff. Thus, your chances of being a carrier (Ff) are two out of three (two-thirds).

If you have a sibling with CF, the probability that you are a carrier is two-thirds. It is not the 1/20 that we would expect in the general population. And your chances of having a CF child would be 2 out of 240 rather than 1 out of 1600 as it is for the general population.

Now, figure out for yourself why the probability of an affected person having an affected child is 1 out of 40.

There are some factors that can change that probability. Most CF males are sterile, and CF females have difficulty conceiving. Thus, the chance that an affected person actually will have an affected child depends on whether pregnancy is possible. It also will depend on the decisions that individuals with CF make about having children. Finally, it depends on how families make use of the recently available carrier and prenatal tests.

---

that the frequency of the heterozygous phenotype (Ff) is almost five percent or 1/20. So, the chance that two unrelated carriers, like my mom and dad, will meet and marry is 1/20 x 1/20 = 1/400. Since each pregnancy between two carriers has a one-fourth chance of producing a child with CF, the frequency of children with CF is 1/400 x 1/4 = 1/16. That gives the birth frequency of kids with CF in the White population."

With my new knowledge of the population genetics of CF, I answered the question, What is the chance that a sibling without CF (brother or sister) of an affected person will have an affected child if he or she marries a person from the general population? The answer is 2/3 x 1/20 x 1/4 = 2/240, or 0.83 percent. And, the chance that an affected person will have an affected child is 1 x 1/20 x 1/2 = 1/40, or 2.5 percent.

I left the Laxalts much wiser than I had arrived. I was impressed with the way they had met a tremendous challenge so successfully.

Sarah had asked me to give her a ride to the basketball game. On the way, we talked. Sarah turned out to be a worrier. She chewed her fingernails and bit her lip. She was not comfortable about being the sister of someone with CF. Did she worry about having affected children in spite of a less than one percent risk?

"Well, yes! Look at my mom and dad. They only had a 1/1600 chance to have an affected child and they had two. My chances are about thirteen times greater than theirs!" Sarah knew that a test for CF carriers had recently become available. She had talked with her parents and Lisa about it because the test sometimes requires the participation of all family members to be most accurate. She also had thought about the prenatal tests available for CF. "What if it turned out I was a carrier and so was my husband?"

"Then I would have to decide whether to end my pregnancy or try to care for a child with CF." Sarah was appalled by the idea of

abortion. She pointed out: "Look at my sister—she gets along okay; she gets all As in school. My baby could turn out like that. I don't think I could have an abortion!"

But has it always been a complete joy to live with her sister? The answer was prompt and vehement: "Absolutely not! I used to be so angry at her because *she* got all the attention. And she never had to help with any of the housework. Because of her we were always broke, and we never could go anywhere or get nice things. I can't get a car, and whenever she gets pneumonia, I get blamed that she caught the infection from me because I had a cold the week before."

Her eyes filled with bitter tears. "My friends are all mad at me because I can't have any parties at my house. Some smoke, or she might catch a cold or the flu from them. And my boyfriend thinks I'm some sort of freak being related to someone with a horrible genetic disease. . . ."

We were almost at the high

school, so I had to ask Sarah what I thought was the big question: "But aren't you glad, Sarah, that you don't have CF?"

At that, she bowed her head, and wiped her tears. She blew her nose and sat quietly for a long time. Then she turned to me and smiled.

"I'm sorry I blew my stack, Alex. I didn't really mean it. You are right, I am pretty lucky! Thanks for the ride." With that, she blew me a kiss, jumped out and joined a group of her friends, laughing. The young man who put his arm around her shoulder did not seem to think of her as a freak.

19. Cloning the CF gene

Since Lisa was born there have been dramatic advances not only in treating people with CF, but also in understanding its underlying cause. In 1985, the gene for CF was localized to a narrow region on chromosome 7. That identification allowed for the possibility of prenatal testing and carrier testing in families that had one child with CF. Then, in 1989, Drs. Lap-Chee Tsui and Francis Collins announced the cloning of the CF gene and the identification of a specific abnormality within that gene that was responsible for 70% of the CF mutations. Among the remaining 30%, believe it or not, more than 600 mutations have been found, most of which occur uniquely in a single family. Nevertheless, a dozen or so are relatively common and, along with the most common mutation, account for about 85% of the mutant alleles.

The product of the CF gene is a protein called cystic fibrosis transmembrane conductance regulator (CFTR). That protein plays a role in the transport of ions—including sodium, chloride, and calcium—in and out of cells. Obviously, its abnormal function accounts for the increased concentration of sodium chloride or ordinary salt in the sweat of an individual with CF. In addition, the disturbed concentration of salts is at least in part, if not totally, responsible for the thickened mucus in the lungs, the bowel and other ducts, as mentioned previously.

Once a gene is characterized and cloned, ap-proaches to gene therapy become feasible. Several research laboratories are engaged in clinical trials designed to insert the normal gene into the respiratory-tract cells (the cells lining the bronchial tubes and their many branches) of CF volunteer patients. One method involves putting the normal CF gene into the genetic material of a harmless virus that is capable of infecting the respiratory tract. When that virus "infects" a cell, it integrates its genetic material into that of its host, and thus, the normal CF gene becomes part of the genome of the infected cells and normal CFTR is produced. That should help thin out or normalize a patient's mucus. Theoretically, the respiratory problem of CF would be cured. Can you think of some problems that would limit the effects of that approach to gene therapy?

The identification of the CF gene now allows for accurate prenatal and carrier testing for CF in families with an affected individual. Furthermore, 70% of carriers (and 50% of affected individuals) in the general population with no family history of CF can also be identified using these tests. Although that provides helpful information to people such as Sarah, it also creates additional ethical, legal, and social dilemmas. For example, should everyone now be tested for CF at birth, or for carrier status at the time of marriage? Would your answer change if we could identify 100% of the CF mutations (instead of the current 85%)? What must you consider to answer that question?

CONSTRUCTING A PEDIGREE

My name is Steve Thacker. I'd like to tell about an experience I had that taught me some important lessons about genetics. For years, my dad, John Thacker, has been telling me stories about his family—his father and mother, his two sisters, and his three brothers. Our family moved from his hometown in Ohio to California when I was only a year old, so I have never really known the people he spoke about. His stories made me curious about all these relatives of mine that I had never seen.

Last summer, I got a chance to satisfy my curiosity. The Thacker family decided to have a family reunion on the 4th of July in the old hometown of Cos Cob, Ohio. Dad could hardly wait to get back and see his family again for the first time in fifteen years. My mother, Marie, is from the same town, and she was eager to see her own family, as well as my dad's. My twelve-year-old identical twin sisters, Laura and Mary Jo, and my nine-year-old brother, Tom, also had heard a lot about the people in Cos Cob. They were delighted at the thought of a trip to Ohio.

We decided to make a real vacation out of our trip back for the reunion. We drove at a leisurely pace, sightseeing and camping along the way. We did so much sightseeing that, on the morning of July 3rd, we realized we were still seven hundred miles from Cos Cob. And the reunion was to begin the next morning at ten o'clock! Dad said we would have to drive straight through and get a good night's sleep so we would

be ready to meet all our relatives and enjoy the reunion. It was a long, fourteen-hour drive. We got into Cos Cob about nine o'clock that evening. Everybody was so tired we decided to get motel rooms and go right to bed.

In spite of all the travel and excitement, I slept pretty well, and so did Tom and the twins. In the morning, after we all got cleaned up, we had a late breakfast. Then we set out for Grandpa and Grandma Thacker's house on Pine Street. We got there at exactly ten o'clock and were amazed to find that we were the last of the Thacker clan to arrive at the reunion. For about fifteen minutes there was a lot of hugging and kissing and squealing. I was introduced to all those aunts, uncles, and cousins that I had heard about but never met. But I have to admit that I didn't get all the names and faces straight.

After the turmoil of the initial meetings was over, I was able to step back and take a better look at my relatives. It was obvious that we all are family. My dad's brother and sister were easy to spot because they look so much like him. My grandparents are older copies of their sons and daughters. The third generation— my sisters, my brother, all our cousins, and I—are alike in many ways. At the same time, all of us have our own special traits that make us different. I remembered one of the big lessons I learned in biology last year. The set of chromosomes we inherit (half from our mothers and the other half from our fathers), in combination

with our environment, is responsible for the similarities and the differences I was observing.

Then I noticed Grandpa Thacker's hands. I thought my Dad had been joking when he used to say that Grandpa Thacker had two little fingers on each hand. But sure enough, it's true. I then checked out Aunt Shirley, Uncle Pat, and Uncle Dave. Again Dad was right! They too have extra little fingers. Aunt Shirley's daughter Sue does not have the extra digits. Uncle Pat has three girls and a boy. One of the girls, Maureen, and the boy, Mike, have the extra fingers. Uncle Dave has a boy, Dan, who does not have the extra digits and a girl, Karen, who does. Dad's sister, Betty, and her son, Jim, do not have the sixth finger. His other brother, Gary, and his two daughters, Patty and Barbara, also do not have the sixth finger. Of course, nobody in my immediate family has the variation.

After my initial observation of this variation in our family, I promptly forgot about it and spent the rest of the day meeting, talking, and playing volleyball with my grandparents, my five aunts and uncles, my ten cousins, and, of course, my own immediate family. The big event of the day was a sit-down dinner that evening. We ate one of the best meals I have ever had. Many stories were told, and each member of the family—all twenty-eight of us—was included in the stories at least once. Before we knew it, it was nearly midnight and time for the reunion to end.

We said good night and went back to our motel, where we talked among ourselves for another hour or so. We spent two more days in Cos Cob visiting my mother and father's old friends. We also fished and picnicked out in the country around Cos Cob.

It wasn't until we were back in California that I remembered the variation I had observed in my grandpa, my aunt and two uncles, and three cousins. I knew that had to be a genetic trait, but I was curious to find out more about it. In my biology class, I had learned that one of the first things you should do to learn about a trait is to make a family history and trace the transmission of the trait through the family. I went to the library and found a chart in a genetics book that showed standard pedigree symbols. Using that information I began to construct our family history. After I had constructed the pedigree, I was able to answer a few questions.

1. Is the trait carried on the autosomes or sex chromosomes?

2. Is it a dominant or recessive trait?

3. Are the people with the trait homozygous or heterozygous?

One question the pedigree did not answer for me was, What is the medical term for the variation of having an extra little finger? For that answer I had to go to the library. When I did, I learned much more about this variation and many other variations common to humans. ▨

LEARNING FROM PEDIGREES

One key to successful genetic counseling is an accurate family history. A good way to illustrate that history is in a pedigree. The dictionary defines pedigree as "an ancestral line; line of descent; lineage; ancestry." Different inherited conditions will show different patterns of heredity in a pedigree. Consider four patterns of heredity:

A. *A dominant condition carried on an autosome (autosomal dominant)*

- Males and females are affected equally.
- Traits do not skip generations (generally).
- A trait is expressed whenever the corresponding gene is present (generally).
- Male-to-male transmission is possible.

B. *A recessive condition carried on an autosome (autosomal recessive)*

- Males and females are affected equally.
- Traits often skip generations.
- Both parents of offspring with the trait are heterozygotes.
- Traits are expressed only in the homozygous state.
- Traits may appear in siblings without appearing in their parents.
- All nonaffected offspring of a parent with the trait are heterozygous carriers of the trait.

C. *A recessive condition carried on the X chromosome (X-linked recessive)*

- Only males show the trait (generally).
- All daughters of a male with the trait are heterozygous carriers.

- Male-to-male transmission is not possible.
- Mothers of males who have the trait are heterozygous carriers.

D. *A dominant condition carried on the X chromosome (X-linked dominant)*

- All daughters of a male with the trait also have the trait.
- Male-to-male transmission is not possible.
- A female with the trait may or may not pass on the trait.
- A female with the trait may have sons or daughters with the trait.

Study the pedigrees provided by your teacher and indicate:

1. the type of inheritance shown by each,
2. your reason for your answer,
3. the phenotype probability,
4. a particular genetic disorder that each case might illustrate. ▨

GUEST EDITORIALS

HUMAN VARIATION

The next time you attend a concert, assembly, or sports event, notice the variation among the people in the crowd. People vary in size and shape of their body parts, shades of skin color, hair styles, behavior, and in many more ways. No two people are exactly alike, even if they resemble each other closely. It's easy to observe a remarkable degree of variation among people.

But, we also must recognize the limits to variation. Only a few people cannot see or hear. Those lacking an arm or a leg are in the minority. None of your classmates weighs ten kilograms or one thousand kilograms, and none is a mermaid or a centaur!

What accounts for the wide range of human variation? How do we explain its limits? One way to explain both variation and its limits is through natural selection. As with populations of organisms in other species, the human population tends to maintain those variations that improve the chances for survival and reproduction. Such variations are adaptive in that they enhance (or, at least, do not reduce) the ability of

humans to deal with their environment. Extreme variations that diverge from what we think of as usual, average, or normal are nonadaptive and decrease the chances for survival and reproduction. Thus, evolution theory helps us understand both the variations and the commonalities we see in people today.

THE LONG AND SHORT OF IT

One need not go to a stadium or a concert hall to observe human variation. Did you know that any time thirty people are in a room together, the chances are very good that two of them will have the same birthday? Also, those thirty people will display both variation in and constraints on some of their characteristics. One obvious and easily measurable characteristic is height. Try the following exercise.

Measure the height of each person in your class. Make a frequency distribution of the measurements, using intervals of five centimeters. Which interval occurs

most frequently? Which occurs least frequently?

Look at the frequency distribution of the heights of a very large group of people, as shown in Figure 1. What is the shape of that distribution? How does that shape compare with the shape of the graph you made? What is the average height of the men whose heights are represented in Figure 1? What is the average for your data?

Now, you should be able to answer the following questions:

1. What factors explain the shape of the distributions in Figure 1 and in the graph you prepared?
2. What accounts for the variation expressed in these distributions?
3. What constrains this variation and prevents people from being very tall or very short?

CAUSES OF VARIATION

What accounts for human height, whether tall or short? To begin with, the genetic blueprint sets limits beyond which it is im-

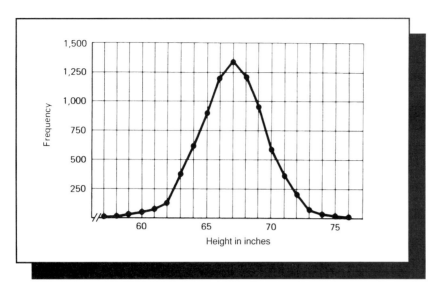

FIGURE 1 ▪ Frequency of heights of 8585 adult men born in Great Britain during the nineteenth century. (From K.D. Hopkins and G.V. Glass. 1978. *Basic Statistics for the Behavioral Sciences.* Englewood Cliffs, New Jersey: Prentice Hall.)

possible to go. The genes are like a building plan. The plan tells the builder what materials to use and just how to use them. If the builder does exactly what is called for, then the only limitation is the plan itself. But, it's unlikely that everything will go just right. The builder might misread the blueprint. Some of the construction workers might make slight changes. The building might go up more quickly or more slowly than planned. Or, the contractors might be unable to obtain the requisite materials, and resort to substitutes. If any of those things happen, the building will diverge from the plan and possibly be imperfect or defective in some way. But, it always will resemble what it was supposed to be—what the blueprint intended.

Who or what constructed the building? The plan? The builder? The construction workers? The materials? The obvious answer is that a blueprint without a builder and materials is just a plan. And a builder who has workers and materials but no plan has nothing to construct. All the elements are necessary.

In human development, the genes are the blueprint. The cellular machinery acts as the builder and the construction crew. The environment supplies the materials. Genes without materials are only a plan. Materials without genes are only chemicals. So, it is idle to argue whether genes are more important than environmental factors, or vice versa.

Which genes determine human height? They are many and varied. They include those genes that govern the chemical processes in the cells of all the organs that contribute to growth and development, especially of the skeleton— the spine, the skull, and the bones that make up the arms and legs. Height, then, is a polygenic quality. Polygenic literally means many genes. Many genes are involved in the determination of height.

These genes require a wide range of nutritional materials, including calcium, phosphorus, sulfur, nitrogen-containing amino acids, and many vitamins. In addition, the growth process requires freedom from infection, adequate rest and exercise, and other favorable conditions. Consequently, human growth is a multifactorial process. Multifactorial means many factors. Many environmental factors influence growth.

If we all had exactly the same genes, diet, and life experiences, we would all be the same height. But, there is great genetic variation among people, as well as differences in diet, in freedom from infection, and in other factors that affect growth and development.

In summary, the height each of us achieves is a product of the genes we inherit from our parents, working in the conditions under which we live. Height is a familial property; children will, on the average, achieve heights similar to those of their parents. When we consider all the combinations that are possible when different genes work under different environmental conditions, it is no wonder that the height of humans varies so much.

We don't have separate, discrete classes of tall people and short people. If we did, a distribution of heights might look like Figure 2, which shows two distinct classes of people with respect to the trait in question. In fact, however, we already have seen that the distribution looks like Figure 3, the distribution for birth weights. What do you think accounts for the differences in the distributions when Figure 2 is compared to Figures 1 and 3?

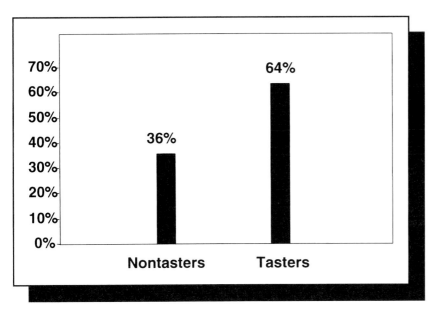

FIGURE 2 ▪ Frequency of tasters and nontasters for phenylthiocarbamide (PTC).

THE LIMITS OF NORMAL

Ask yourself where in the distribution tallness ends and shortness begins. How tall is too tall? How short is too short? How can one determine what is normal or abnormal? Suppose you were asked to choose two points that mark the limits of medium height in the distribution. Which points would you choose? How might you explain or defend your choices?

Look again at the distribution of heights in Figure 1. Why are so many people represented by the measurements in the middle and so few by those at the ends? It must be that more genes contribute to middle heights than to the two ends; and those that contribute to the extremes are rare. The contribution from the conditions of life are likely to be distributed similarly. Thus, the very tall and the very short stand out by virtue of their infrequency, as well as their height. This rarity may itself add to the difficulty of living in a society organized by and for people of more average height. Doorways, beds, stairs, automobiles, and clothing are seldom created with people of extreme height in mind. Remember that tallness and shortness are not separate qualities, but only infrequent parts of a continuous distribution of measurements.

THE HEIGHT OF INDIVIDUALS

So far we have suggested that variations in height among human beings are due to both genetic and environmental factors, and that height is both a polygenic and a multifactorial property. But, how can we answer the more specific question of what accounts for an individual's height? For those people near the middle of the curve—those that all would agree are normal in height—we cannot single out any particular gene and say that it contributed so much or detracted by so much. Nor can we single out any environmental conditions, except chronic infections or episodes of nutritional deprivation.

The more height diverges from the average, however, the more

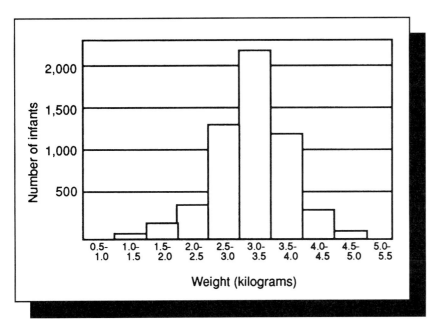

FIGURE 3 ▪ Birth weights of infants (surviving, normal, male and female combined). (From Van Valen, Leigh, and Gilbert W. Mellin. 1967. "Selection in Natural Populations: 7. New York Babies [Fetal Life Study]." *Annals of Human Genetics* 31:109–27.)

likely we can point to some gene, chromosome, or developmental abnormality as responsible for excessive or retarded growth. Some abnormalities due to hormone deficiencies of the thyroid or pituitary might be treatable. Others are due to single-gene defects that involve proteins associated with bone growth and development. Some chromosomal abnormalities are associated with slowed growth as well. At the opposite pole, unusually tall stature is characteristic of Marfan syndrome and acromegaly (caused by overproduction of the growth hormone) and the XXY, XXX, and XYY chromosomal constitutions.

As we have seen, infections and nutritional deficiencies can interfere with growth, either because the deficiency of some vitamin or nutrient has been so complete as to stop growth, no matter what the genotype, or because the deficiency has interfered with the growth of some genetically vulnerable people, while sparing others. The recognition of extreme differences in stature might be important, then, as a way of identifying the conditions that caused them. Such conditions may be correctable or treatable, and the ge-

netic contributions may be important for genetic counseling.

Many of the genes associated with alterations in growth produce other adverse phenotypic effects. It is easy to identify these effects in families and, therefore, easy to establish modes of inheritance. But, there are other, multifactorial causes of growth alteration for which neither the genes nor the environmental events or agents can be precisely described. Therefore, the greater the divergence in height from the average, the more likely we are able to detect the genes and the environmental events responsible. But some cases, even at the extreme ends of the distribution, continue to defy our current knowledge. Nevertheless, we can be sure that genetic variation is involved.

SUMMARY

In this section, we've dealt with some subtle and important principles. We can state them as six propositions:

1. Neither genes nor environmental events are sufficient to produce any phenotype. Both are necessary.

2. The way a person's genes work in the environment during the period of growth determines the person's height.

3. Measurements of stature are continuously distributed, so it is difficult to decide just where medium height ends and short or tall stature begins.

4. Individuals who fall outside the normal range for stature can be evaluated for causes that can be treated or prevented. Remember that such people are not abnormal; only their height is outside the normal range.

5. Extreme tallness and shortness are the exceptions, rather than the rule. That is because of the low frequencies of the genes and environmental factors involved; those genes and environmental factors likely have been selected against during human evolution.

6. It is possible to describe the genes and environmental events involved only for the most divergent and most nonadaptive cases. That is a property not of the genes themselves, but of the limitations of the knowledge currently at our disposal.

PROGRESS MADE ON NEUROFIBROMATOSIS

Probably 100,000 Americans have a genetic disorder called neurofibromatosis. There is no cure. And the social and vocational rejection many victims of this disorder encounter may hurt more than the complex medical problems they must endure.

A former college teacher talks matter-of-factly about the prob-

lems "we uglies" have in getting jobs and promotions. A young woman in her twenties wonders if "anyone will ever love me." The mother of a six-year-old fights for public understanding and research essential to her son's fu-

---

*Adapted from Joan Beck, Copyrighted © 1981, *Chicago Tribune*. Used with permission.

ture. Perhaps this column* can help pull neurofibromatosis out of this cruel closet, with the aid of one of its victims, Frances Zapatka, a former teacher.

OUT OF THE CRUEL CLOSET

Neurofibromatosis is marked chiefly by skin discolorations—

called café-au-lait spots (coffee with milk)—and by the uncontrollable growth of tumors on the nerves just under the skin—sometimes masses of little tumors, sometimes large ones. Tumors also can appear on the auditory nerves, causing deafness; on the optic nerves, causing blindness; and on the spinal cord and in the brain, where they can be fatal. Sometimes the bones enlarge and become misshapen and the spine develops a severe curvature. About fifteen percent of the people who have neurofibromatosis (NF) have learning problems of variable severity.

We now can prevent severe disfigurement with plastic surgery, although the tumors often return and surgery sometimes seems to make them grow even faster. Zapatka has already had 130 operations and "of course" faces more. She says tumors are growing in her upper palate, spreading and lifting the bone beside her eyes. It's painful, she acknowledges, but "the pain is not constant."

Neurofibromatosis is one of the most common of all hereditary disorders and is caused by a single, abnormal gene. It is inherited as an autosomal dominant trait, so there is a fifty percent chance in each pregnancy that a baby will inherit the disorder if even one parent has it (Figure 1). But, symptoms vary greatly and occasionally don't start until middle age, after a parent has already unwittingly passed the disorder on to a new generation.

NO EXPLANATION FOR SOME

Nearly half of the cases appear to be new, spontaneous mutations, for which doctors have no explanation. There are at least two forms of neurofibromatosis. One is called the central form (or NF$_2$), and it most frequently causes tumors of the central nervous system, especially the nerves for hearing. The other, peripheral neurofibromatosis or NF$_1$, is more common, with tumors forming under the skin and variable numbers of café-au-lait spots.

The genes for both forms have been mapped (NF$_1$ to chromosome 17 and NF$_2$ to chromosome 22), cloned and studied, with most of the work concentrated on the more common NF$_1$. The product of the NF$_1$ gene is called neurofibromin, and it is an enzyme whose function is still unknown. Clearly it has an influence on perhaps several nerve growth factors and the symptoms of the condition appear to be due to a deficient amount of neurofibromin being produced by the single mutant allele; the normal allele cannot compensate. Perhaps not surprising in view of the fact that the NF$_1$ gene is very large and the symptoms of the condition are so variable, nearly every patient and family studied has a different mutation; some have deletions, others have point mutations that cause single amino acid substitutions in the enzyme protein that diminish its activity. And, as we are discovering with surprising frequency now that we have the technology to distinguish the parental origin of chromosomes, the NF$_1$ gene shows imprinting. There is a marked preference for the new mutation cases to occur in the paternally derived NF$_1$ allele (see "Imprinting," page 104, for more details). At present, we don't understand the reason for that preference.

The existence of so many different mutant alleles that cause NF$_1$ has made prenatal diagnosis by direct gene analysis difficult for most families at this time. In families where there are more than two affected individuals, however, linkage studies are often possible. As we see for many genetic disorders, especially dominant ones, the marked variability of symptoms, even within families, makes

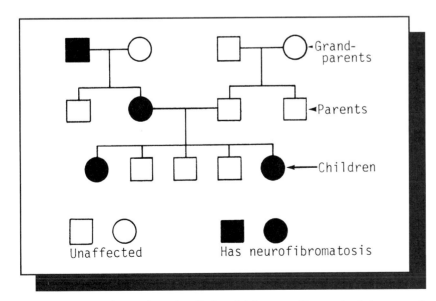

FIGURE 1 ■ Pedigree for a family in which neurofibromatosis is present.

the ethical considerations extremely difficult. The majority of individuals with NF_1, perhaps seventy-five to eighty percent, have café-au-lait spots and possibly a few lumps under the skin; they never suffer from the more serious complications described earlier.

"My mother had four children before she discovered she had neurofibromatosis," Zapatka says. "She still has only minor symptoms." But Frances, the oldest child, began developing signs of the disease shortly after birth when tumors started to grow around her nose and "along the smile lines of my cheeks," and her eyelids began to droop. Doctors diagnosed neurofibromatosis when she was two years old. She was treated first with radium and chemical injections, which were ineffective, and had the first of those 130 operations when she was eight.

A younger sister has much milder symptoms, little more than tiny, acne-like tumors. "Otherwise, she is very attractive, as I probably would have been," Frances comments. Three younger brothers have apparently beaten the genetic odds.

Tumors or not, Zapatka has had an energetic and productive life. After her father deserted the family when she was eleven, Frances took over considerable responsibility for her younger brothers and sister while her mother went to work. Frances worked her way through college and graduate school and taught for fifteen years at elementary, high school, and junior college levels, often taking tough classes no one else would handle. But when her medical problems increased and medical bills became almost as large as her income, she was forced onto disability.

FOUNDATION SET UP

Neurofibromatosis victims and families have gone public on the disorder. In 1978, they set up the National Neurofibromatosis Foundation, Inc. (95 Pine Street, 16th Floor, New York, NY 10005, 1-800-323-7938), which encourages research, helps families, and fosters understanding.

In fact, there are foundations, associations, and support groups in the U.S. and Canada for most genetic disorders, no matter how rare. They are easy to find through organizations such as the National Society of Genetic Counselors and the Alliance of Genetic Support Groups. Information on many genetic support groups is available on the Internet (see "Be an Active Participant," page 80).

It takes a while to learn to cope with the pubic, Zapatka has found. "Most people experience uneasiness on meeting those with the form of neurofibromatosis I have. Some may be frightened. Some may consider it (us) obscene, leprous, or contagious. Some see only our disfigurement. Others get to know us and enjoy us as persons." There has been some progress, then, both in society's acceptance of the victims of this disorder, and in scientists' understanding of it. But not yet enough of either.

Basic Genetics

Why Are There Two Kinds of Cell Division?

You came into being as a single cell—a fertilized egg or zygote that was the result of the union of an egg cell from your mother and a sperm cell from your father. That single cell contained your genome, your hereditary material (DNA), half from your father and half from your mother. Through cell division, that one cell eventually became trillions as you grew. Even after you stop growing, new cells will replace old cells, dead cells, or cells that have lost their ability to function properly. Each new cell produced must have a complete complement of DNA identical to the DNA in that first cell that was you. Mitosis is the type of cell division that makes it possible for a single parent cell to form two genetically identical cells.

Mitosis is a cyclical process: the products of one division recycle into yet another division. The events that occur at any one moment in mitosis blend imperceptibly into those that occur at the next moment. Pictures and diagrams of mitosis in books are the stages of the process and are like the still frames of a videotape.

Think of mitosis as the videotape and not the still frames.

Our discussion starts at the stage called interphase, which is not considered a stage of mitosis. For years biologists referred to interphase as the resting stage because under a light microscope you could see things happening during mitosis, but during interphase nothing seemed to be happening. Today we know that this is not correct. During interphase the chromosomes are long and slender, like threads. The genes on these chromosomes are exerting their maximum influence. There is a period of protein synthesis and then a period when the replication of DNA takes place, which results in paired identical chromosomes. Then, there is a second period of protein synthesis. It might help you to understand interphase mitosis if you review the appendices "The DNA Molecule" and "The Genetic Code and Production of Protein" in this book. What would the status of a cell be at the end of interphase?

If you watch a living cell under a microscope during prophase, the chromosomes become visible.

That is because the long, slender, thread-like structures become progressively shorter and thicker. If the normal number of chromosomes in a human cell is forty-six, how many chromosomes would you see at this time? Think back to what happened to the chromosomes during interphase.

As we continue to watch our cell under the microscope, we see the replicated chromosomes begin to take a position down the center of the cell. A still frame at this point is called metaphase.

When the lining up is complete, the replicated chromosomes will pull away from each other and one of each pair moves to the opposite sides of the cell. A still frame at this point is called anaphase.

When the chromosomes reach the sides of the cell, they begin to become long and slender again. Mitosis ends with this final stage called telophase. Then the cytosol (cytoplasm) will pinch together and divide, forming two cells. How many chromosomes will be in each cell? What else will be identical in the two new cells? The new cells are identical to the parent cell, which of course does not exist anymore. Mitosis occurs in all body or somatic cells. In what cells does it not occur? The end results of trillions of mitotic cell divisions are growth, repair, and replacement of the body's cells.

One type of very special cells goes through a process of cell division that is different from mitosis. These are the reproductive cells, the eggs formed in the ovaries of females and the sperm formed in the testes of males. Before you read further, think of reasons why mature eggs and sperm cannot be produced by mitosis. Write your reasons in your notes and evaluate your ideas when you finish this activity.

Meiosis is limited to cell division in special cells in the testes and ovaries. As in mitosis, meiosis is cyclical. For the purpose of this discussion, we will describe meiosis in stages. But keep in mind that meiosis, like mitosis, is like a videotape and the stages are still frames of that tape.

DNA replication takes place during interphase I. There is no protein synthesis as there was in mitosis. Because of that, what will not happen to this cell?

The chromosomes become visible during prophase I. We see that not only have the forty-six chromosomes made duplicate copies of themselves, each pair has joined with its similar pair forming a structure called a tetrad (four chromosomes joined together). How many tetrads would you see?

The cycle continues, and during metaphase I the tetrads line up along the center of the cell. In anaphase I the tetrads are pulled apart in pairs. The cell now di-vides to form two cells in the stage called telophase I. How many chromosomes will be in each cell?

Interphase II begins with protein synthesis, but with no DNA replication. What will happen to the cell? What won't happen to the chromosomes? The cycle continues and metaphase II is similar to metaphase in mitosis in that the paired chromosomes line up in single file along the center of the cell. If you looked through a microscope at a cell going through mitosis and one going through meiosis, what major difference would you see? At anaphase II the paired chromosomes pull apart and the chromosomes move to the sides of the cells. Cell division takes place in telophase II, resulting in four cells (called gametes). How many chromosomes will each gamete contain?

Earlier we asked you to write in your notes reasons why eggs and sperms could not be produced by mitosis. The most important reason you could have given was that they would have forty-six chromosomes in them. Fertilization of an egg by a sperm would then result in a cell with ninety-two chromosomes. The next generation would have 184. The next 368. Obviously, that is not possible, so eggs and sperm must be produced by a process that will reduce the chromosome number by one-half. Fertilization brings the number back to the normal forty-six, the number characteristic of *Homo sapiens*.

A simple way to describe mitosis and its results is as follows: DNA replication is followed by cell division, followed by DNA replication, followed by cell divi-sion, and so on. That ensures a continuity in number and the genetic code in all the somatic cells of an individual.

1. Develop a similar description for meiosis and its results.
2. A class of twenty-five biology students scanned the entire length of an onion root tip on a microscope slide. The students observed all five stages of mitosis and counted the number of cells in each stage. There were between 240 and 300 cells in the interphase stage. In the other four stages—prophase, metaphase, anaphase, and telophase—there was a total of twelve to fifteen cells. What do those data suggest about the length of time for each stage?
3. There are approximately two hundred inherited disorders in which cancer has been reported as a regular or occasional feature. Although we don't completely understand the causes of cancer, we think most result from a combination of genetic and environmental factors. Identify a genetic factor that could result in cancer.
4. The nuclei of cells of organisms such as a rat, a radish, and a human undergo mitosis before cell division. What is the basic significance of mitosis?
5. Before scientists had discovered and observed meiosis, they had predicted the occurrence of this type of cell division. How was it possible to predict that meiosis occurred before there was actual evidence for it?

CHANCE, PROBABILITY, AND GENETICS

We learned earlier that pairs of replicated chromosomes (a tetrad) separate during anaphase I of meiosis and move to opposite poles of a cell. In humans, that means that twenty-three paired chromosomes are moving in one of two directions. What controls which paired chromosomes will migrate with any other paired chromosomes? What determines which one of the 350 million sperm ejaculated during intercourse will fertilize an egg? What decides that a man who carries the gene for a lethal, recessive disorder will marry a woman who carries the same gene? The answer to those questions is chance. The causes that determine the outcome of those events are numerous and unidentified. So, we call those and many other events in genetics chance events.

Chance is not a vague or nebulous concept. We can express it quantitatively by its probability of occurrence. If we know two things about a chance event, we can calculate the probability that a chance event of choice will occur. We must know how many equally probable events are possible, and the number of these that will give us our desired event. We can put these two quantities into a formula for calculating probability.

p(probability) =

$$\frac{\text{number of events of choice}}{\text{number of equally likely events}}$$

We can use the formula to determine the probability that expectant parents will have a girl. The number of events of choice is one (a girl), and the number of equally likely events is two (a boy or a girl). By using the formula, we can determine that the probability is $1/2$, or .5, or 50 percent.

This formula for calculating the probability of a chance event is an important tool in human genetics. Mathematicians have developed three laws of probability that also are valuable tools in understanding genetic events.

First Law of Probability: The results of one trial of a chance event do not affect the results of later trials of the same event.

Assume that a couple has had seven male children in a row. What is the probability that their eighth child will be a girl? The first law of probability tells us that the births of the previous seven boys were all independent events that have no effect on the eighth child. We may intuitively feel that a girl is next, but probability has no memory.

Second Law of Probability: The probability that two or more independent events will occur together is the product of their probabilities of occurring separately.

Two or more events are said to be independent when the occurrence of any one of them does not affect the occurrence of any of the others. Use the second law to determine the probability that a couple will have two boys. Remember, each conception is an independent event.

Third Law of Probability: The probability that either of two mutually exclusive events will occur is the sum of their separate probabilities.

We say that events are mutually exclusive when the occurrence of any one of them excludes the occurrence of the others. What, for example, is the probability that a three-child family will have two girls and one boy? There are three ways to get that combination of children:

1. girl-girl-boy
2. girl-boy-girl
3. boy-girl-girl

We use the second law to determine the probability for each of the three possibilities.

1. $p = 1/2 \times 1/2 \times 1/2 = 1/8$
2. $p = 1/2 \times 1/2 \times 1/2 = 1/8$
3. $p = 1/2 \times 1/2 \times 1/2 = 1/8$

Each of the three possibilities is mutually exclusive. (That is, you can't have family one or three if you have a girl-boy-girl combination.) Therefore, we use the third law of probability to get the final answer, which is what?

If you understand the laws of probability and the methods for calculating the probability of chance events, you will understand human genetics more easily.

THE GENETICS OF HUMAN POPULATIONS

Up to this point, most of the articles in this magazine have been on the genetics of individuals and families. We can apply Mendel's laws and other principles of genetics to the entire human population or to subsets of the human population. Geneticists ask many questions about populations of individuals, without regard to their precise identity. For example:

- In the United States, what proportion of the population has type AB blood?
- How many "harmful" genes (such as the ones responsible for cystic fibrosis or Huntington disease) are there in a population?

Geneticists want to know how frequently certain genes occur in a population. The study of gene frequencies is the basis of population genetics.

A population geneticist who wants to know a particular gene frequency must first determine the specific population she wants to study. For our discussion, she will select all students of high school age in Milwaukee, Wisconsin. Because that population is so large—about sixty thousand—the geneticist will study it by taking a random sample of the population. Instead of sixty thousand students, she will select five thousand, and that group will become her population to study.

Next, the geneticist counts the individuals in this sample who possess the variation under study. Let us say that the geneticist is interested in the number of high school students in Milwaukee who can taste a chemical substance named phenylthiocarbamide (PTC). Some students (called tasters) taste it easily, but others (called nontasters) cannot taste it at all. The ability to taste PTC is a dominant trait, symbolized by the allele T. Those who cannot taste PTC carry only the allele t, which symbolizes the recessive trait. What are the genotypes of the tasters and the nontasters?

The geneticist and her co-workers contact five thousand high school students and test whether they can taste PTC. The geneticist finds that 3200 students can taste PTC, and 1800 students cannot. She determines the percentage of tasters by dividing 3200 by 5000, which results in 0.64, or sixty-four percent. By a similar calculation (1800/5000), she finds that the nontasters make up 0.36, or thirty-six percent of the population. What are the genotypes of the tasters and nontasters? The thirty-six percent that are nontasters represent thirty-six percent of the population with a tt genotype. But the sixty-four percent who are tasters could have a TT or a Tt genotype. What percentage have what genotype?

The geneticist working on the PTC problem thinks of all the genes that the members of a population might contribute to the next generation as the gene pool. In this investigation, a certain proportion of the alleles in the population are Ts and the rest are ts. To determine those proportions, the geneticist uses a mathematical formula called the Hardy-Weinberg principle, developed in 1908 by G. H. Hardy, an English mathematician, and W. Weinberg, a German physician.

Before continuing our investigation, we must understand the derivation of the Hardy-Weinberg principle:

1. In the gene pool, p equals the proportion of alleles for the dominant trait.
2. q equals the proportion of alleles for the recessive trait.
3. $p + q = 1$ (1 represents all—100 percent—of the alleles in the gene pool).
4. therefore, $p = 1 - q$.
5. and, $q = 1 - p$.
6. Each gene pool is derived from two separate gene pools—the male gene pool in the sperm and the female gene pool in the egg.
7. As a result of fertilization, the two gene pools combine into one.
8. The second law of probability is applied: $(p + q)(p + q) = p^2 + 2pq + q^2$.
9. Resulting in the Hardy-Weinberg principle: $p^2 + 2pq + q^2 = 1$.

We now will apply the Hardy-Weinberg principle to our study.

1. Relate the Hardy-Weinberg values to the genotypes in our study:
   $p^2 = TT$
   $2pq = Tt$
   $q^2 = tt$
2. Given: a population in which sixty-four percent are tasters and thirty-six percent are nontasters.

3. What value in the Hardy-Weinberg principle do the nontasters represent? $q^2 = .36$
4. What two values can we determine from that information? $q = .6$ and $p = 1 - .6 = .4$
5. We now have the gene frequencies in our gene pool. $p = T = .4$ and $q = t = .6$
6. What are the genotype frequencies in this gene pool?
   $p^2 = TT = (.4)^2 = .16$
   $2pq = Tt = 2(.4)(.6) = .48$
   $q^2 = tt = (.6)^2 = .36$

These gene and genotype frequencies will remain constant from generation to generation unless one or more of the following occur:

Mutation. Alleles involved in the study mutate.

Natural selection. One of the alleles provides a reproductive advantage, that is, the people who have that allele produce more offspring.

Nonrandom mating. If people begin to select mates on the basis of the trait involved, the allele frequencies will change in future generations.

Emigration and immigration. The movement of people into or out of the population will affect the gene frequencies.

Population genetics has become a valuable tool for geneticists. The more we know about the distribution of alleles that affects the health and abilities of the human population, the more success we are likely to have in coping with the problems in which the welfare of the entire population is involved.

RECESSIVE TRAITS AND THE GENE POOL

According to the Hardy-Weinberg principle, once a population reaches a state of equilibrium, gene and genotype frequencies will remain constant from generation to generation. At first glance, that does not make sense. Shouldn't the frequency of the allele for a recessive trait decrease from generation to generation? Let's state that question as a hypothesis: If recessive traits are expressed only in homozygous individuals, then the frequency of the allele for that trait will decrease from generation to generation. You and a partner can test that hypothesis by building a model population.

Materials

For each team of two students:
pencil
1 copy of Worksheet 1
80 red pop beads
120 yellow pop beads
box
calculator

Procedure

In the model population in "The Genetics of Human Populations," 36% of the people cannot taste PTC. They have the genotype tt. The ability to taste PTC is a dominant trait, represented by T.

1. Recall what you have learned about the Hardy-Weinberg principle. What is the frequency of the t allele? What is the frequency of the T allele?
2. Use pairs of pop beads to construct a model of that population. Use the red beads to represent the T allele and yellow beads to represent the t allele. Make 16 pairs of two red beads (homozygous tasters, TT); 48 pairs of one red and one yellow bead (heterozygous tasters, Tt); and 36 pairs of two yellow beads (homozygous nontasters, tt). Your total model population should equal 100 pairs.
3. Place these 100 pairs of beads in a box and mix them thoroughly. Have one person at a time withdraw, at random, two pairs of beads. These pairs represent matings. Record each mating on Worksheet 1 in the column marked Number of Crosses.
4. Repeat the mating procedure 50 times. Return the beads to the box after each mating.
5. Assume that each pair of "parents" produces four "offspring" and that the genotypes of these four progeny are those that are theoretically possible (T.P.) in single gene inheritance. For example, in the TT x TT mating, the theoretically possible offspring are 100% TT, 0% Tt, and 0 percent tt. Tally the offspring from each of the 50 matings. Next, total the number of TT, Tt, and tt offspring.
6. Calculate the genotype frequencies of the offspring. Then calculate the frequencies of the alleles in the gene pool of the offspring.

1. How do the genotype and allele frequencies of the offspring in your matings compare with the genotype and allele frequencies of the parents? Discuss and explain your results.

2. Do these data support the hypothesis?

3. Why are physical and mathematical models such as this one important tools in biology?

4. In this population, no factors cause the allele frequencies to change. Such factors include mutation, natural selection, isolation, emigration, and immigration. What would have happened to the allele frequencies if one or more of those factors had been operating? Why?

A GENE POOL IN A STATE OF CHANGE

In the activity "Recessive Traits and the Gene Pool" (page 41), you learned what happens to the frequencies of genes and genotypes when all genotypes have an equal chance to reproduce. What would happen to the frequencies of tasters and nontasters if individuals with one or the other genotype had a greater or lesser chance of survival and reproduction?

Materials

For each team of two students:
data from Worksheet 1 ("Recessive Traits" activity)
4 copies of Worksheet 2
1 copy of Worksheet 3
pencil
pairs of red and yellow pop beads ("Recessive Traits" activity)
box
calculator

Procedure

1. The original population in the activity "Recessive Traits and the Gene Pool" (which we shall symbolize by G_0) had the following genotype frequencies: TT = .16; Tt = .48; and tt = .36. The allele frequencies were T = .4 and t = .6. When you made 50 crosses with this G_0 population and calculated the genotypes of the 200 offspring (the G_1 population), you probably found little, if any, difference in the frequencies.

2. Now, you will introduce a new factor into the original population. Suppose that a spontaneous mutation makes all the homozygous recessives in the G_0 population sterile. Determine the effect of that mutation on the G_1 population. Use the same number of population crosses as in the "Recessive Traits" activity. Transfer the numbers from Worksheet 1 ("Recessive Traits" activity) to the new table in Worksheet 2. Label the figure generation number G_1. Calculate the number of offspring and their expected genotypes. (Your totals will be different this time because of the spontaneous mutation that took place.)

3. After you have determined the genotypes of the offspring in the G_1 generation, calculate the genotype and allele frequencies and record your answers in the space provided. Use the genotype frequencies to determine the number of individuals in this new generation and use them as a new breeding population. Label a fresh copy of Worksheet 2 the G_2 generation.

4. Place these 100 pairs of beads in a box and mix them thoroughly. Have one person at a time withdraw, at random, two pairs of beads. These pairs represent matings. Record each mating on the G_2 copy of Worksheet 2 in the column marked Number of Crosses. Repeat the mating procedure 50 times. Return the beads to the box after each mating.

5. After the 50 matings, assume that each pair of "parents," except those that are sterile, produces four "offspring" and that the genotypes of these four progeny are those that are theoretically possible in single gene inheritance.

6. Calculate the genotype and gene frequencies of this G_2 generation.

7. Use these genotype frequencies to determine the number of individuals in the next gen-

eration (G_3) to be used as the breeding population.

8. Repeat this procedure until you have generated data for four generations of offspring (G_1 through G_4).

9. Summarize the genotype and gene frequencies for the G_1 to G_4 generations on your copy of Worksheet 3.

INTERPRETING THE RESULTS

1. What is happening to the frequency of the allele for the recessive trait?

2. In which genotype in G_4 do you find most of the alleles for the recessive trait?

3. How does that compare to G_0?

4. Compare your data from "Recessive Traits and the Gene Pool" (the G_0 generation) to the data in this activity. Is natural selection enough to explain the differences? Explain your answer.

5. Even under extreme natural selection, can the allele for a recessive trait be eliminated from a gene pool? Why?

SO MANY ANTIBODIES—SO FEW GENES

Your body is constantly battling microscopic and submicroscopic foreign invaders. Your immune system is the defense. It distinguishes between self and nonself. Your intact skin and mucous membranes help to keep infectious organisms, such as bacteria and viruses, out of your body. Those are nonspecific parts of the immune system. That is, each responds the same way to all potential invaders. Those foreign substances that manage to overcome a nonspecific barrier encounter a second line of defense: antibodies. The antibody response to foreign invaders is much more specific than the nonspecific response. In fact, the immune system produces a different type of antibody for each foreign invader or antigen.

Antibodies are proteins, which means that genes control their production. Immunologists estimate the human immune system can produce about one million different antibody proteins. Geneticists estimate that there are about eighty thousand genes in the human genome. Do you detect a problem with those figures?

This investigation will help you explore the mechanism by which the human genome encodes information for such an incredible amount of antibody diversity. But, you will need some background information first. Refer to Figure 1 as you read the following information.

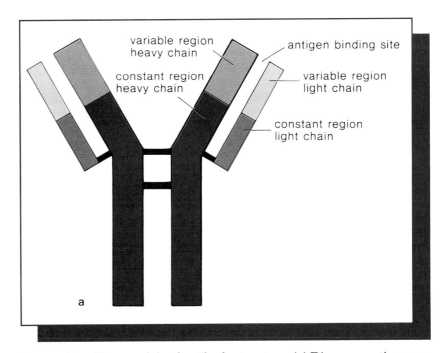

FIGURE 1 ■ Two models of antibody structure. (a) Diagrammatic representation showing the variable and constant regions of the light and heavy chains and the identical halves of the molecule. (b) A computer-generated model that gives a sense of the three-dimensional shape of an antibody molecule. The variable regions of one light and one heavy chain combine to form an antigen-binding site; each antibody molecule has two antigen-binding sites.

1. Each antibody molecule consists of four chains of amino acids: two identical light chains and two identical heavy chains.
2. One light chain and one heavy chain combine to make half of an antibody molecule.
3. In each half, one end of the heavy chain and one end of the light chain together provide one antigen binding site. That is where the antigen will bind to the antibody. Once an antigen is bound to an antibody, it can no longer harm the tissues or organs of the body. How many binding sites does each antibody molecule have?
4. Remember that antibodies are proteins, which are made from amino acids. For different antibody molecules, the sequence of amino acids varies greatly at the end that forms the antigen binding sites. This part of the molecule is called the variable (V) region. At the other end, the sequence of amino acids is very similar. This is called the constant (C) region. A thick region joins the variable and constant region together. This is called the joining (J) region. What does the highly variable

end of the antibody molecule explain?

Materials

For each team of two students: paper clips: 1 red, 1 pink, 1 orange, 1 yellow, 1 blue, 1 green, 1 white, 1 black

Procedure

1. Assume that your body has been invaded by 12 different infectious organisms and must produce a different antibody light chain against each one. The complete gene for a light chain includes three types of smaller DNA segments: variable (V); joining (J); and constant (C).
2. Working as a team, use your paper clips to construct a complete sequence of DNA that will code for the production of an antibody light chain. Hook the paper clips together in the sequence indicated in Figure 2, using the following key: R (red), P (pink), O (orange), Y (yellow), Bl (blue), G (green), W (white), B (black).

Each colored paper clip represents a different type of DNA segment. The segments are labeled V_1

through V_4 for the variable region genes and J_1 through J_3 for the joining region genes. There is only one C gene.

3. Each complete light chain gene must have one V segment, one J segment, and the C segment. Construct a gene sequence for a light chain by combining V_1 (red), J_1 (blue), and C (black).
4. *Return* V_1, J_1, and C to the original DNA strand. Now combine V_2, J_2, and C.

Before continuing, answer the following questions.

1. How do the chains constructed in steps 3 and 4 differ?
2. Is there enough genetic information present in these two sequences (V_1, J_1, and C; V_2, J_2, and C) to produce ten distinct light chains? How many chains are possible?
3. How many light chains could you produce using V_1 and each of the J segments?
4. Assume that any V segment can combine with any J segment (and C) to produce a complete gene for an antibody light chain. Use the paper clips to construct complete genes that will code for 10 distinct light chains. Record

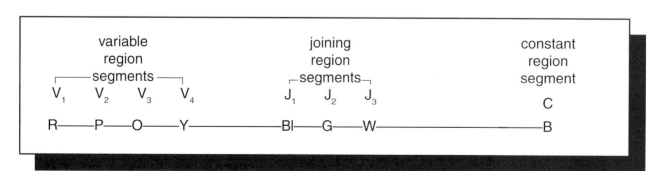

FIGURE 2 ▪ Paper clip sequence.

the sequences. You already have completed two:

1   R   Bl   B V_1 J_1 C
2   P   G   B V_2 J_2 C

Now work with your partner to answer questions 5 through 10.

5. How many complete light chains could you produce using the genetic information in the original strand?

6. If there were approximately 250 V segments, 5 J segments, and 1 C segment for an antibody light chain, how many different light chains could that information code for?

7. Using the symbols V, J, and C, complete the following, general equation to show the number of light chains that can be produced by any given stretch of DNA: number of light chains = ? x ? x ?

8. We can define a gene as a sequence of nucleotides that codes for a specific product, such as tRNA, an enzyme, or a structural protein. Does the model for the production of antibodies support that definition? Why or why not?

9. Traditionally, a gene was considered to be a continuous stretch of DNA, all of which was translated into protein. Does the model for the production of antibodies support that definition? Why or why not?

10. How is the structure of the antibody molecule related to its function?

ESSAYS:
Hard Choices

THINKING ABOUT ETHICAL QUESTIONS

Making decisions about right and wrong is often very difficult. First, one must identify what *can* be done. Usually, more than one choice is available. If one choice is obviously superior to the other alternatives, then the decision is made with relative ease and confidence. But many times two or more choices seem defensible and, worse yet, both may seem good solutions to our problems. A genuine dilemma arises when one can identify and understand logically acceptable sets of reasons that seem to justify opposite courses of action.

A respected philosopher, Gerald Dworkin, has written about that problem.* He points out that there is no "theory of morality" that corresponds to theories in the physical and life sciences. For example, the theory of evolution in biology (a) explains many observable facts, (b) predicts the outcomes of experiments,

and (c) allows scientists to generate new hypotheses that can be investigated by using the methods of science. We can accept or reject hypotheses based on the experimental results. Moral questions cannot be approached in that way. We can't conduct an experiment to show that one idea is more legitimate than another.

But the lack of a testable theory of ethics does not mean that we can't answer moral questions, or that one point of view is necessarily just as good as another. Dworkin explains that there are three broad categories of moral considerations: goals, rights, and duties. Applying those categories to the analysis of ethical problems does not guarantee easy answers. But Dworkin's definitions and examples do give us a way to think through difficult ethical questions. His approach does so by helping us to identify what counts as clear-headed reasoning and what does not.

Goals, rights, and duties may be in conflict in many situations. We become confused when we fail to distinguish one from another. If we look at the arguments that

people give to justify one course of action over another, we often find all three. Usually, however, one of them—a goal, a right, or a duty—will be offered as the most important or most compelling reason. Even specialists in the field of ethics disagree about which is the most important in any given situation. So, we should not be surprised when we find ourselves torn between competing positions when we must make a difficult decision.

First, let's examine what we mean by goals, rights, and duties. Then we can go on to look at ways in which these ideas can be applied to the analysis of ethical problems.

Goals. One can judge the morality of an act by looking at what it intends to accomplish. We can ask ourselves about the objectives or outcomes. We can focus on consequences. In that pattern of thought, we might judge a "good" outcome as morally correct regardless of how the goal was achieved. That may be termed the any-means-to-an-end idea.

Consider some examples. Suppose we could agree that it is

---

*Dworkin, Gerald. "Analyzing Ethical Problems." *Hard Choices*, A Magazine on the Ethics of Sickness and Health. Boston, Massachusetts: Office of Radio and Television Learning.

morally desirable for a father to care for his children, and that a father should take any and all necessary steps to assure that his children survive. We might then conclude that stealing food from a neighbor to feed his starving children would be a morally justified choice. We might be tempted to qualify that judgment with statements like "if there were no other way," but in offering such exceptions we would call on other moral ideas such as the duty to be honest.

Another example may help further define moral arguments that rely on goals. A physician may view his or her primary mission as the preservation of life. Life, by that view, is to be sustained as long as possible by any means possible. In this goal-oriented view, then, a physician might refuse to disconnect a respirator on a comatose patient whose vital signs are maintained only by the machine. The physician might take such action against the wishes of the patient's family or even against the previously expressed wishes of the patient. In the physician's view, going against the rights of the patient would be justified by the *goal* of preservation of life.

Rights. Moral arguments that involve the notion of rights are familiar in many areas besides genetics, health, or medicine. One is said to have a *right* if one is entitled to a certain kind of treatment, no matter what the consequences. The founders of our country spoke of the inalienable rights of life, liberty, and the pursuit of happiness. They claimed that one could expect those things regardless of the situation or the outcomes—just by virtue of being an individual human being. Thus,

we have derived as part of our political heritage the belief that each person has a right of free speech, a right to vote, or a right to own property.

It is not hard to think of situations in genetics and medicine where one might use arguments about rights to support a particular course of action. For example, consider the matter of informed consent. Informed consent—the full disclosure of all relevant information so a patient can accept or refuse some treatment—rests on the belief that a person has a right to know all there is to know and to make up his or her own mind freely. In that view, no treatment can be given unless the person understands the procedure and chooses to undergo it.

This right to know and to choose is widely acknowledged among physicians, and informed consent is practiced routinely. Imagine, however, a physician who feels that some treatment will save a patient's life. But the physician knows also that this patient will refuse if all the possible consequences of the treatment are spelled out. How much information should the doctor share? Would the physician be justified in violating the patient's *rights* because the *goal* is to save the patient's life.

You can readily see how goals and rights can come into conflict. What situations can you imagine in which the rights of different individuals are in conflict? How are rights gained or lost? Can rights be limited by authority or by mutual agreement?

Duties. The duty or obligation to act in a certain way is often cited in ethical arguments. We typically think we have a duty to tell the truth, to keep a promise,

or to help a friend. Usually, duties are justified by suggesting that the act will achieve some worthy goal, or that the act is required because of someone's right. Thus, we can derive duties from goals or rights that may be in conflict with either goals or rights or even other duties.

A defending lawyer's relationship with the accused lawbreaker involves duties that are derived from goals and rights. It is the *right* of the accused to have legal representation, regardless of innocence or guilt. It is the attorney's *duty* to represent the accused, also without regard to guilt. Further, it is the lawyer's *duty* to seek the goal of acquittal, without consideration of the consequences for either the defendant or for society.

The logic of duties can be very complicated. Suppose a dying man asks a physician not to take any extraordinary steps to prolong his life. Does the doctor have a *duty* to comply with the man's wishes because of the patient's *right* to die? Or, does the physician have a *duty* to ignore the man's wishes because of the *goal* of the preservation of life?

Obviously, ethical dilemmas usually involve all three elements: goals, rights, and duties. It is seldom possible, however, to combine all three into a single, satisfying solution. Why, then, are they so important? Those notions help us *understand* situations that involve difficult choices. Goals, rights, and duties won't make our choices for us. But identifying the competing goals, rights, and duties that seem to be operating in a particular situation will help us see more clearly the choices that are available. It may even sharpen the focus enough to allow us to predict benefits and losses with greater precision. Also, goals, rights, and duties are the language of justification, of defending the responsibleness of one's choices.

We can analyze case studies (see "Case Studies in Genetics," page 49) using goals, rights, and duties as tools to work through a practical problem. Consider the following case:

Leonard Steinman is a twenty-three-year-old college student of Jewish descent. In a large university-sponsored screening program, he is identified as a carrier of Tay-Sachs disease. During counseling, he reveals that he is engaged to marry a Jewish woman in three months. He tells the counselor that he has a twenty-year-old sister and a sixteen-year-old brother. He refuses to reveal his carrier status to his fiancée, his parents, or his brother or sister. He also forbids the counselor to tell them.

What should the counselor do? The counselor has a number of alternatives; she can attempt to persuade Leonard, refer Leonard to another counselor, or wait for Leonard to change his mind. But, for the sake of argument, let's consider the most extreme and opposite choices: to tell or not to tell. The counselor can justify going against Leonard's wishes on several grounds—such as the *rights* of the others to know about their own health status, the *goal* of health promotion for all these individuals, and, thus, the counselor's *duty* to pursue that goal. On the other extreme is the counselor's *duty* to maintain confidentiality and respect Leonard's wishes. We can see that *duty* as derived from Leonard's *right* to privacy or his *right* to determine how he will interact with other family members. In keeping silent, the counselor would see the *goal* of respect for persons, in this case Leonard, as of a higher value than the *goal* of promoting health in the possible future offspring that Leonard or other members of his family might have.

To expand this analysis further, consider the ethical arguments if Leonard's fiancée called and asked the counselor about the results of Leonard's screening. Should the counselor divulge the results to her? Suppose the situation were even more complicated. What if the fiancée indicated that she herself is a carrier? What should the counselor do then? Why?

The identification of goals, rights, and duties doesn't solve the problem, but, it does make the choices clearer. And, it can break one big decision down into several smaller ones. Sometimes, we can resolve smaller decisions one at a time and make the situation more manageable. For example, in the case of Leonard Steinman, we can identify at least three rights that may be in conflict. Leonard has rights. His fiancée has rights. And the child they may have someday will have rights, too. One can imagine landmark court cases in which decisions might go against or in favor of any one of those three. Depending on the number and nature of such cases, society might come to accept that one of these three has rights that predominate over those of the other two. Once the question of rights has some history of debate and decision, we might see a trend emerge that would make the decision to tell or not to tell a bit easier in the future.

Nevertheless, we are not likely to see some theory of morality arise that will settle these questions once and for all, at least not in the near future. Students of ethics are still some way from a single unifying theory that will serve philosophers as the theory of relativity serves astrophysicists. As Dworkin noted: "In many ways it is easier to agree on the nature of distant galaxies than it is to agree about the proper way to treat one's neighbors."

CASE STUDIES IN GENETICS

Many human problems result from new knowledge and technologies in genetics. By far, the majority are ethical problems. Ethics describes systems of moral behavior—what individuals, or groups, believe to be right or wrong conduct. Philosophers who specialize in ethics examine the different arguments that justify one course of action over another. In simplest terms, those arguments relate to issues of goals, rights, and duties as outlined in "Thinking about Ethical Questions" on page 46.

One of the tools for analyzing ethical problems is the case study, a story of a real, or possibly real, situation. Case studies in ethics are stories in which one can readily see conflict between two or more positions or two or more courses of action.

The case studies that follow present three such situations. The questions that follow each case study will help you think through the alternative courses of action and the arguments that would support each course. There are no right answers, but that does not mean we can simply propose solutions in the absence of sound analysis. Each situation requires you to analyze personal and professional codes of conduct. We don't expect that any agreement can, or even should, be reached. Instead, discuss each case in terms of what different individuals might consider as the relevant goals, rights, or duties.

Case study 1. Marilyn Parks, twenty-three, is referred to a genetic counselor. She has given birth to a second child with physical defects and mental retardation. A physician has diagnosed both children as victims of fetal alcohol syndrome—a set of developmental defects caused by a mother's excessive drinking during pregnancy. Marilyn tells the counselor that she does not plan to have any more children.

The counselor learns from the physician that Marilyn is psychologically fragile and has been severely depressed since the birth of her second child. In addition, the physician relates that Marilyn's home environment is not supportive. Her husband has threatened to leave her because he thinks she is not a good mother.

1. Should the counselor tell Marilyn what caused the defects in her two children? Why or why not?
2. Suppose that, instead of planning no more children, Marilyn revealed that she is three months pregnant with her third child. What should the counselor do? Why?

Case study 2. Pat Jackson is a young Black man. He has been screened for the sickle-cell gene and has been found to be a carrier. He has no health problems and no symptoms of the disorder. He applies for health insurance. He does not know where this particular company stands on policies for those with sickle-cell trait. The application asks, "Do you have any genetic or inherited disorders?"

1. How should Pat answer the question? Why?
2. What are the possible consequences of the action you recommend?

Case study 3. Your community advertises a screening program for carriers of sickle-cell disease. You know that the disorder can be severe, even fatal in some homozygotes, while relatively mild and treatable in others. A local newspaper accuses the screening program of promoting discrimination and racism through an attempt to discourage reproduction among Blacks.

1. What would you want to know before you could decide whether there is any truth to the paper's accusations. (We are not asking you to decide whether the paper is correct. Instead, we want you to reserve judgment and identify what a person should try to find out before making up his or her mind.)

DISCOVERY OF THE GENE FOR HUNTINGTON DISEASE

In the fall of 1983, scientists announced a remarkable breakthrough in research on Huntington disease (HD), a serious, degenerative disorder that generally has its onset between the ages of thirty-five and forty-five. The scientists, James F. Gusella, Harvard Medical School; Nancy S. Wexler, Hereditary Disease Foundation; P. Michael Conneally, Indiana University; and their colleagues announced that they had found a genetic marker for HD on chromosome 4. That discovery—and the subsequent isolation, cloning, and characterization of the gene— has allowed many people at risk for HD to find out whether they carry the gene and will therefore develop the disorder. The availability of such information raises many ethical dilemmas, some of which we will examine in the activity "Who Should Be Told?" But first, let's look at these remarkable discoveries.

Researchers have known for a long time that HD is caused by a defect in a single gene. The disorder is inherited as an autosomal dominant trait. No one knows what causes this fatal disorder, and we have made little progress on treatment. A person who has HD has a fifty percent chance of passing the gene for the disorder to each of his or her offspring. The problem is complicated by the late onset of the disorder. Many people already have had children by the time they find out they have HD. Therefore, their children have a fifty percent probability of developing the disorder themselves.

In genetics, we use markers to map genes to specific chromosomes. Sometimes a marker can even help map the gene to a specific region of a chromosome. A marker might be an easily detectable enzyme or a fragment of DNA. When the marker was found in 1983, scientists were hopeful that they soon would isolate the gene itself. But that search took another ten years. During that decade, scientists studied large numbers of families using DNA markers located at varying distances from the HD gene.

Some individuals at risk of inheriting the HD gene, but who had not yet shown any symptoms, wanted to know whether they had inherited the gene. The family was tested, and the individual was given a risk figure based on linkage. Before being tested, obviously, the risk was fifty-fifty. After the linkage studies, they were given a probability of having or not having the HD gene depending on how close the marker useful in their specific family was to the HD gene. For example, a person might learn that the chance of having inherited the HD gene from an affected parent was less than five percent. Sometimes, however, the best figure a laboratory might come up with, based on linkage, was, say, less than twenty-five percent. That leaves a large margin for error in a condition that has severe symptoms. Furthermore, some at-risk couples wanted to use linkage analysis for prenatal diagnosis and might elect to terminate a pregnancy if a "probably affected" fetus were

found. But considering termination with a twenty-five percent chance that the fetus did *not* have the HD gene was of great concern. We discuss other important issues relevant to that in "Who Should Be Told?"

THE HUNTINGTON DISEASE GENE

In 1993, biologists succeeded in isolating and cloning the HD gene itself. The results of this work were a surprise to many. The mutant gene turned out to be another example of gene expansion due to trinucleotide repeats. (For more information, see "Anticipation: Some Genes Do Change," page 105.) The reason for the surprise was that HD pedigrees do not, in general, show anticipation (progressively worsening symptoms with earlier onset of disease as the gene passes from generation to generation). There is one exception—when the HD gene is passed to a child of either sex from the father, the age of onset tends to be earlier and in the very rare instances of childhood onset HD, it is invariably the father who has transmitted the HD gene.

At the molecular level, HD is one of the allelic expansion diseases with small-scale instability. The trinucleotide repeat number increases two to three fold over normal size (6–37 CAG repeats in normal people; 35–121 in individuals with HD). The expanded sequence, in contrast to what we see in the large-scale changes of conditions such as fragile X syndrome and myotonic dystrophy, is relatively stable as the gene passes from generation to generation, except for a greater tendency to expand when the father is the affected parent. Thus, in addition to some degree of anticipation, the HD gene sometimes shows imprinting, a difference in expression depending on the parental origin of the mutation (see "Imprinting," page 104).

Now that we know the actual mutation for HD and testing is available to families, identification of people with the HD gene has reached nearly one hundred percent accuracy and we no longer use linkage. In fact, the direct DNA analysis of the gene is less difficult and even less expensive than the family studies required for linkage analysis. Many people at risk for HD who were given fairly high levels of error are being retested and in some cases have had their HD gene status reversed. For example, a person given an eighty-five percent chance of having the HD gene based on linkage could be told, after the direct gene test, that there is close to a one hundred percent chance that the HD gene is *not* present.

At this time, the function of the HD gene is not known nor is the nature of its product. We hope continuing research will sort out the nature of the mutation and provide clues about the prevention of symptoms in people identified as having the gene but who do not yet suffer the consequences.

WHO SHOULD BE TOLD?

The ability to determine who has the HD gene and who does not has created a new set of ethical dilemmas. You will analyze them with the help of your teacher. Among the problems that arise is whether to test individuals and inform them that they carry the gene when so little can be done to help them. There is no treatment or cure for HD; the disorder simply gets worse for ten to twenty years before the patient dies.

Is it better for a person to live with the knowledge that he or she has a fifty percent chance of developing the disorder, or to learn that the disorder will develop? Some people claim that knowing that the disorder will develop "deprives the patient of hope and thus increases the risk of severe depression." Others say that certainty is preferable to uncertainty. Certainty allows the individual to make informed decisions about marriage, careers, having children, and financial planning. In addition, they add, the ethical principle of autonomy, one's right to make decisions about one's life, argues in favor of providing such information. The availability of psychological counseling and support from social services also plays a role in whether to inform people.

The researchers who discovered the marker—and subsequently, the gene—were concerned about the predictive nature of their work. In fact, after the original linkage studies were completed, Gusella and his colleagues did not inform the members of the American family whose cells contain the marker that the HD gene is present. Nor did they publish all the information about the Venezuelan family where the marker also was present. In their research paper they stated: "Although a number of younger at-risk individuals were also analyzed as part of this study, for the sake of these family members the data are not shown, due to their predictive nature. The data are available upon request if confidentiality can be assured."

Information derived about one individual, of course, often has implications for other individuals. The researchers themselves pose the following, hypothetical situation:

I am a twenty-five-year-old woman, and my husband and I would like to have a baby. My mother is forty-five and at risk. Her father had HD. My mother doesn't want to know if she is going to get HD because she's well adjusted to being at risk and already has her family. But I want to know.

Obviously, if the twenty-five-year-old woman is found to carry the gene for HD, she inherited it from her grandfather through her mother. Her mother, therefore, has the gene for HD and will develop the disorder. Would it be possible for the younger woman to maintain her silence in that situation?

And would her actions give away the results—if she adopted a baby, for example?

It also is possible to tell whether a fetus has the HD gene. By using amniocentesis or chorionic villus sampling (see "Prenatal Diagnosis," page 70), scientists can examine the cells from the fetus and determine whether the fetus has inherited the HD gene from the parent who is known to have it.

This approach raises controversial and complex issues related to reproductive decisions. Some couples might decide to continue the pregnancy if the fetus were found to have inherited the HD gene. Others might choose to abort the fetus. Those problems are especially difficult for disorders that have a late onset, such as HD. Should a pregnancy be terminated when a child that would result from the pregnancy would not develop the disorder for perhaps thirty-five years or more? Many great contributions to society were made by people who were considerably younger than thirty-five. On the other hand, is it morally defensible knowingly to bring into the world a child with a fatal genetic disorder? The discovery of the HD gene adds yet another complication to the situation because it introduces the real possibility of gene therapy that could prevent the symptoms in people who have the mutation and who are detected early. Ultimately, the parents have to make the decision, based on their own values. A genetic counselor can help them explore the options, but the counselor cannot and should not decide for the couple.

As with all such issues, there are no clear-cut, right or wrong answers. But, we must all be informed about such issues, so that we can participate effectively in the development of the policies that will guide the use of new technology. These issues are examples of the problems that arise as a result of rapid progress in biomedical science.

IN THE NEWS

ROBERT VANDENBERG WINS SPECIAL OLYMPICS

Jefferson High School in Spanishburg, West Virginia, has had special education classes for almost five years. One of the special education students is Robert Vandenberg. Robert, age sixteen, recently took part in the Special Olympics. He won two first-place medals—one in the long jump and one in swimming.

Robert was thrilled. He wore his heavy medals on a ribbon around his neck when we went to his home for an interview. Robert explained that, because his mom and dad had to work, his Uncle Joe had taken him on the train to Roanoke, Virginia. Joe had rented a car in Roanoke. They stayed in a motel near the sports arena with almost two hundred other young people from the region who were there to compete in many different events. The motel had a gym and a swimming pool, so Bob was able to practice. To keep in shape, he and his uncle ran twice a day. After the Olympics were over and Bob had collected his medals, Joe took him to the zoo, several movies, and a hockey game.

During our interview, we noticed that Bob's speech was some-times slurred and was difficult to understand when he became excited. We also noticed some other unusual aspects of his appearance and behavior. So, we decided to interview Bob's parents to find out more about him.

Harold and Margaret Vandenberg were happy to talk to us—mostly, they said, because so many people misunderstand Bob and his condition. Bob is not stupid, but his view of life and the way he expresses himself are certainly less mature than one might expect in a sixteen-year-old boy. Bob has a jolly personality, an enjoyment of life that is gratifying to his family and friends. The fact that he is so active in sports evokes admiration from other students in his school. But not everyone is kind. Especially when he was younger, Bob was the brunt of many cruel jokes and insults.

Bob is busy at home. While his mom and dad both work in offices, he does most of the housework, including laundry and cleaning. He cuts grass, rakes leaves, and shovels snow. He takes care of the family's two beautiful Irish setters. He washes the car cheerfully, but he is unhappy about being denied the chance for driver's education.

Bob also does most of the grocery shopping. Harold or Margaret orders meat by phone and Bob picks it up. Since Bob cannot read very well, Harold has made a small ring-bound notebook of photographs of grocery items. To make up the grocery list, Harold puts into the notebook pictures of the items needed that week. To fill the order, Bob simply turns to the pictures in the book. Bob cannot make change, but all the checkout clerks know him, and he always brings home the correct amount.

In his vocational training classes, Bob has learned machining, assembling, and packing skills. After graduation, he will be able to work full time, under supervised conditions. He is looking forward to his first job.

Bob often goes to a nearby town on the bus with some of his friends. When he goes on these trips, he wears a bracelet that gives his name, address, phone number, and the statement, "I have Down syndrome."

What is Down syndrome? The

Vandenbergs explained that Down syndrome (DS) is named after John Langdon Down, a British physician, who gave one of the first descriptions of the condition. Down noted that the condition was common in institutions for the mentally retarded. Indeed, DS is the most common form of severe mental retardation in the population: one-third of people who are mentally retarded have DS. The proportion would be higher, but the death rate of DS infants is high. DS infants often die as a result of serious heart defects. Only about half of DS children reach adulthood.

What kind of a condition is DS? It is a genetic syndrome—specifically a malformation syndrome. That means that the affected person has several malformations—more than we would expect based on the frequency of these malformations in the population. Mostly, the malformations are minor, but they help doctors diagnose the condition. Some of these minor malformations occur in individuals in the general population as normal developmental variants.

The physical appearance of people with DS is a result of several subtle developmental changes. People with DS usually are shorter than others of the same age. They usually are heavier and their muscles are not as firm. Their posture may not be good, and they frequently have flat feet. The facial features of a person with DS may include a sloping fold of the eyelid; small, low-set ears; low nasal bridge; a broad neck; and an open mouth and protruding tongue.

People with DS tend to be gentle and happy. We noticed those characteristics when we spoke with Bob. The I.Q. ranges from twenty to sixty, with a median of forty to fifty. Those at the upper level of the I.Q. range sometimes can learn to read and write. A few can learn to get along in the community. The vast majority are unable to function on their own in society. Some are profoundly retarded and need institutional or nursing home care.

At one time, many DS children were automatically admitted to institutions for the mentally retarded shortly after birth. Today, the exact opposite is true. Most parents try to educate the child to attain maximum skill and knowledge. That is why we see more people with DS in public, doing many of the things that everyone else does. They are living much happier and more productive lives than ever before. Bob's pride in his Olympics medals, his freedom of movement, his ability to perform useful work, and his anticipation of future semi-independent work and living conditions have given him hope and optimism. He has achieved a level of dignity and self-respect that, until recently, was rarely possible for retarded people.

Women thirty-five or older have a greater risk of having a child with DS than do younger mothers. And the risk increases with each additional year of maternal age. But three-fourths of the children who are born with DS have mothers under the age of thirty-five. Why, then, do we say the risk is greater for older women? Women thirty-five and older have a disproportionately large number of children with DS. Of all children born, fewer than ten percent are born to women thirty-five and older. But about twenty percent of the children with DS are born to older moth-ers. For women under thirty-five, the risk of having a child with DS is approximately 9 in 10,000 (0.09 percent). For women thirty-five and older, the risk is approximately 7 in 1000 (0.7 percent). So, the risk is almost eight times greater than one would expect on the basis of random chance in the population.

If you have already tried "Countdown on Chromosomes" on page 57, you probably have a good idea of the genetic causes of the kind of DS that Robert has. Extra chromosomal material causes DS, usually an entire extra chromosome 21. That is called trisomy 21. Robert has forty-seven chromosomes instead of the usual forty-six (Figure 1).

Another, more rare cause of DS, occurs when most of an extra chromosome 21 is attached to another chromosome, such as 14 or 22. Those people with DS have the normal number of chromosomes—forty-six—but on closer inspection, one can see that the extra 21 chromosome is attached to 14 or 22, giving the characteristic number for DS of forty-seven.

How does the extra chromosome get into the fertilized egg? It does so through a process called nondisjunction, which occurs during meiosis, when ova and sperm are forming. Nondisjunction may occur in the first or second meiotic division. Nondisjunction means that during cell division, a pair of chromosomes does not separate. Instead, both chromosomes enter a single cell. That results in one egg or sperm that has only twenty-two chromosomes. Such reproductive cells usually die. Another cell has twenty-four chromosomes. It contains two members of the pair. During fertilization of such a cell, a third

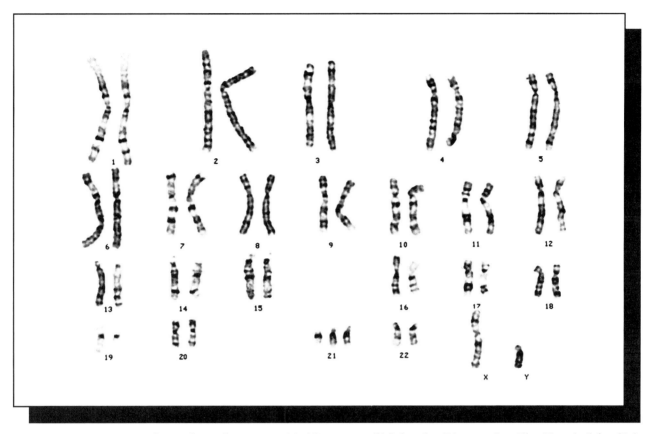

FIGURE 1 ▪ Karyotype for male with trisomy 21. (Courtesy of Molecular Diagnostics, The Hospital for Sick Children, Toronto, Canada.)

chromosome of the same kind is added. Trisomy results.

Nondisjunction is quite common in humans. It is possible that as many as half of all human embryos are chromosomally abnormal. All human chromosomes have been seen in a trisomic state. The most common occurrence is trisomy of chromosome 16. Most humans with trisomies die. Fewer than half of the humans with trisomy 21, a rare case of trisomy 13, or trisomy 18 survive. About sixty percent of all spontaneously aborted human embryos and fetuses have a chromosomal abnormality. About six percent of still-born infants have a chromosomal defect. Of live-born infants, only 0.6 percent show the effects of trisomy. Table 1 shows the situation at birth.

Those data make more sense if we review what karyotypes show. The twenty-third pair of chromosomes is the sex chromosomes. They occur in two forms, called X and Y. The sex chromosomes determine the sex of an individual.

An individual with two X chromosomes is female. And an individual with one X and one Y is a male. The other twenty-two pairs of chromosomes are called autosomes. Autosomes are all the chromosomes except sex chromo-

TABLE 1 Types and frequencies of chromosomal abnormalities at birth.		
Total infants examined	54,749	
Total abnormal	326	(0.6%)
Type of abnormality	**Number**	**Percentage of total**
Sex chromosomal	120	0.22%
Autosomal trisomy	74	0.14%
Structural defects in chromosomes	132	0.24%
Total	326	0.6%

somes. Thus, trisomy 21 is an autosomal condition. Autosomal defects usually are very serious. They are almost always associated with severe or profound mental retardation and a high death rate.

Theoretically, we can detect all of the more obvious chromosomal defects before birth using amniocentesis or chorionic villus biopsy (see "Prenatal Diagnosis," page 70). But, right now, it is neither possible nor desirable to perform prenatal diagnosis on every pregnant woman. Genetic counselors usually offer prenatal diagnosis to women over thirty-five or to those with a history of a genetic disorder in their families.

STUDYING HUMAN CHROMOSOMES

Not many years ago, we didn't know the correct number of human chromosomes. We hadn't invented the simple ways of studying chromosomes inside cells that we use now. We studied chromosomes by slicing cells very thinly and trying to count the pieces of chromosome material in each section. That was like trying to assemble a very difficult puzzle, and the process often produced incorrect answers. Fortunately, there are new methods that give us reliable results. We now study chromosomes by growing cells in a test tube in a broth of special chemicals and nutrients. We then break open the cells and examine them under a microscope.

To study chromosomes, we must first obtain cells that will grow. We can use many different kinds of cells, but white blood cells and skin cells are the most commonly used because they are simple to obtain and easy to grow. Why not use red blood cells?

After a few white blood cells or a small bit of skin have been obtained, we place the tissue in the special nutrient mixture. Different types of cells have slightly different nutritional needs, and it's important to select the right materials in the correct amounts. Furthermore, we must keep the fluid at just the right temperature and it must have enough oxygen. We have to be careful to keep bacteria and other microorganisms out or they too will grow. The cells take from several days to several weeks to begin growing. Sometimes chemicals are added to stimulate growth and cell division.

Once the cells begin to grow and divide, we add other chemicals to prevent normal cell division. Cell division usually is stopped when the chromosomes are tightly condensed and lined up in the center of the cell.

After the cells reach that stage, we add other chemicals to cause the cells to swell with fluid. The nuclear membrane bursts, and the chromosomes from each cell nucleus fall in a small cluster. We then stain the chromosomes so they are visible under a microscope. Specially trained laboratory technicians look carefully at the chromosomes of many different cells. They take pictures through the microscope of "well-spread" clusters where they can see all the different chromosomes. Technicians develop the picture, cut out the chromosomes, and sort them according to their size, shape, and banding pattern.

It's possible to use several different types of staining chemicals to reveal more about the minute details of chromosome structure. That process is tedious and time consuming, with the result that chromosome studies are often rather expensive—as much as $600 at present in most laboratories. Imagine how difficult and expensive it can be to study an entire family. Nevertheless, in special situations, such investigations can provide extremely valuable information about the health of individuals and about their reproductive risks.

COUNTDOWN ON CHROMOSOMES

Understanding human genetics requires some study of chromosomes. Technicians in hospitals and genetic service centers use special techniques to take pictures of the chromosomes in a dividing white blood cell. (We can't use red blood cells because they do not have nuclei.) The picture looks like Figure 1. We then cut the chromosomes out of the picture and arrange them in a standard sequence called a karyotype (see Figure 2). Note that there are twenty-three pairs of chromosomes in this karyotype. Twenty-two of the pairs are called autosomes and are numbered from one to twenty-two. The twenty-third pair is the sex chromosomes. There are usually two X chromosomes in a female karyotype and an X chromosome and a Y chromosome in a male karyotype. What is the sex of the person whose chromosomes we see in Figure 2?

Geneticists have established standards for identifying each of the forty-six human chromosomes. Each pair of the twenty-two autosomes has been numbered, from one to twenty-two, according to length. The XX of the female and XY of the male are designated as the twenty-third pair. Even though it is difficult to see in Figure 2, each chromosome is double, because replication already has occurred prior to the separation of the chromosomes during mitosis.

It is difficult to arrange chromosome clusters exactly according to number; in Table 1, however, we arranged the twenty-three pairs into seven groups according to

TABLE 1 Chromosomes grouped according to size and location of the centromere.		
Group	Chromosomes	Characteristics
A	1–3	Very long with metacentric (median) centromeres
B	4 + 5	Long with submetacentric (set off from the middle) centromeres
C	6–12 + X	Medium length with submatacentric centromeres
D	13–15	Medium length with acrocentric (at or very near to end) centromeres
E	16–18	Somewhat short with submetacentric centromeres
F	19 + 20	Short with metacentric centromeres
G	21 + 22 + Y	Very short with acrocentric centromeres

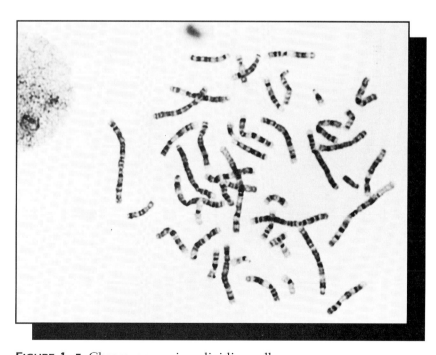

FIGURE 1 ▪ Chromosomes in a dividing cell.

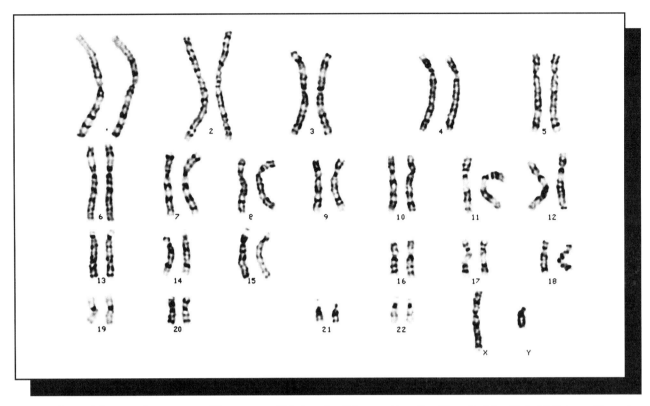

FIGURE 2 ▪ A human karyotype. (Courtesy of Molecular Diagnostics, the Hospital for Sick Children, Toronto, Canada.)

size and location of the centromere. You will receive two pictures of chromosome clusters. Use the information in Table 1 to make a karyotype of each cluster on a karyotyping form.

Materials

1 copy of Worksheet 1
1 copy of Worksheet 2
2 copies of Worksheet 3
scissors
pencil
paper
tape or glue

Procedure

1. Work first with the copy of Worksheet 1; then repeat the steps for Worksheet 2.
2. Circle each chromosome with a pencil.
3. Cut out the individual chromosomes.
4. Arrange the cut-out chromosomes in pairs and decide to which of the seven groups they belong.
5. Use glue or tape to attach each chromosome in its proper place on Worksheet 3, the karyotyping form.

INTERPRETING THE KARYOTYPES

1. What is the sex of the individual whose chromosomes appear on Worksheet 1? on Worksheet 2?
2. Compare the two karyotypes you made. What specific difference can you find?
3. How important is that difference? If you don't already know, you should read "Robert Vandenberg Wins Special Olympics" on page 53.

SICKLE-CELL TRAIT IS NOT SICKLE-CELL DISEASE

Emmet Richardson went to a neighborhood health fair sponsored by a local television station and the medical society. The screenings, demonstrations, and exhibits related to health promotion and disease prevention impressed Emmet. He had his blood pressure taken at one booth, and his lung capacity determined at another. His hearing was excellent, but he couldn't see the tiny letters near the bottom of the eye chart. Emmet already knew he needed glasses. Next, he stopped at a booth marked "Hemoglobin-opathies" (HE muh glow bun AH puh thies). He watched as others ahead of him had their fingers poked and several drops of blood placed on a card. Emmet read the brochures. He watched and listened to the automatic slide presentation.

The narrator talked mostly about sickle-cell disease. Emmet knew something about sickle cell. Michael Patterson, a member of Emmet's church, had the disorder and was often tired and suffered periodic pains in his joints.

Emmet learned that sickle-cell disease is an inherited disorder of red blood cells. The disorder causes the red blood cells to become distorted into shapes resembling sickles (Figure 1) when the oxygen level in the blood is low. Because the cells that become sickled cannot flow easily through the tiny capillaries, they create a "traffic jam." That decreases the blood supply to vital organs, such as the heart, spleen, kidneys, and brain, which can damage those organs.

Although the symptoms of sickle-cell disease vary, there are some general clinical features. Infants may experience jaundice (yellowing of skin and other tissues due to the breakdown-products of red blood cells), anemia, pain, and a predisposition to infection. In later years, due to the traffic jam in the capillaries, the disorder may cause leg ulcers, anemia, kidney failure, stroke, and heart failure.

Emmet realized why Michael was short of breath, tired, and could not participate in the YMCA basketball league. Emmet also learned that not all patients with sickle-cell disease have the clinical symptoms. Some people may be completely free of serious illness, and others may have only a few symptoms. The severity of sickle-cell disease and its symptoms differs for various age groups. While some people with sickle-cell disease do not have any problems, others die at an early age.

The narrator of the slide show used a pedigree to explain how sickle-cell anemia is inherited (Figure 2). Emmet saw that in the first generation, each parent has one gene for normal and one gene for sickle hemoglobin. Sickle hemoglobin is a recessive characteristic. Sickle-cell disease—the disorder—occurs only when a person has two genes for sickle hemoglobin. A person with one normal gene and one sickle gene has sickle-cell trait. Both parents shown in the pedigree have sickle-cell trait. People with sickle-cell trait do not have the disorder. Their life expectancy is the same as that for people with normal hemoglobin. Only under unique environmental conditions, such as severe oxygen depletion, might people with the trait exhibit some

---

Editor's note: A few years ago, our editorial office received a number of reports that some companies were denying insurance to certain individuals. These people did not have sickle-cell disease but, instead, they had sickle-cell trait. It was obvious that there was confusion about the difference between sickle-cell disease and sickle-cell trait. We published several articles to explain the situation and to clear up some of the confusion.

Our efforts—as well as those of other magazines, newspapers, television and radio stations, teachers, and volunteer groups—have helped alleviate the problem. We do not receive as many reports of misunderstanding or outright discrimination as we once did. Nevertheless, we know that the problem is not solved completely. Some people still are confused about what sickle-cell disease is and how it is inherited.

One of the reporters in our education division recently submitted a story that seemed timely and important enough to include in our IN THE NEWS section. We hope our readers will learn as much from the story of Emmet Richardson as we did.

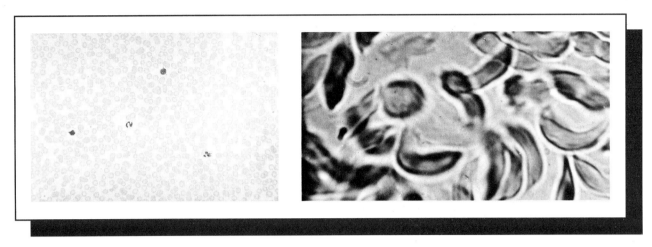

FIGURE 1 ▪ Normal cells and sickled cells.

sickling. (Some examples are heavy exercise at very high altitudes, deep sea diving with a depleted oxygen supply, and riding in an airplane at high altitude and the cabin depressurizes. Of course, we would all be in trouble if we were involved in the last two examples.)

Emmet examined the pedigree carefully. The pedigree shows a couple with five children. The first two inherited normal hemoglobin genes from both parents. The third

and fourth are carriers; they have sickle-cell trait, like their parents. The fifth child received the sickle hemoglobin gene from both parents. She has sickle-cell disease. When both parents have the trait, the probability that any given offspring will inherit two sickle hemoglobin genes is twenty-five percent (Figure 3). That percentage does not mean that twenty-five percent of the children in one family will automatically have sickle-cell disease. What does it

mean? Probabilities are based on large numbers of families with carrier parents.

Emmet saw that no one was in line. He decided to find out if he was a carrier. The technician asked Emmet to fill out a form and to have one of his parents sign at the bottom. Emmet knew that his parents would not be home from work for another half hour. He decided to stay and ask more questions about sickle-cell disease.

Emmet noticed that the drops of blood on the cards were drying. He asked the technician where the cards were going and what would happen to them. The technician told him that the cards would be taken to the University Hospital. There, other technicians would analyze the DNA for the mutation associated with sickle hemoglobin. A person who does not have sickle-cell disease or sickle-cell trait will not show the mutation in any of his or her DNA. A person with sickle-cell trait has some DNA with the mutation and some that doesn't. A person who has

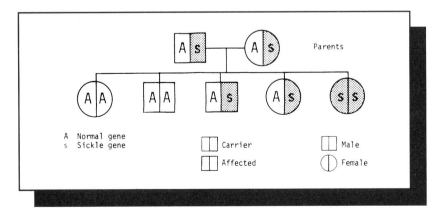

FIGURE 2 ▪ Inheritance of sickle-cell disease.

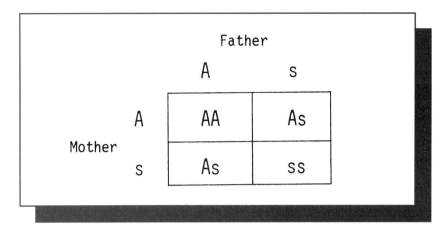

Father

	A	s
A	AA	As
s	As	ss

(Mother — rows labeled A and s)

FIGURE 3 ▪ When both parents have the trait, the probability of sickle-cell disease in the offspring is twenty-five percent.

sickle-cell disease has only DNA that has the mutation. Emmet was surprised to learn that there are about two hundred other known variant forms of hemoglobin. Each form is caused by a different gene mutation, but only a few cause health problems.

Emmet looked at his watch. It was about time for his parents to be home. He thanked the technician and left the health fair with the consent form. Later, Emmet returned with his parents and two younger sisters, who had decided to be screened for sickle-cell trait. The technician assured them that the test results would be confiden-

tial and that they would be notified in a week.

The results of the DNA analysis showed that Emmet, his father, and one of his sisters were carriers. Emmet's first concern was that he was less than healthy because of the sickle gene he had as a part of his genome. The genetic counselor assured him that he was very healthy. She explained that all of us carry dozens, and possibly hundreds, of poorly functioning or even nonfunctional genes that make us susceptible to a variety of things under certain circumstances. She suggested that Emmet get a copy of the article

"Why Study Human Genetics?" (which is in your magazine on page 4) to help him better understand that he is not unique when it comes to having deleterious genes.

Then the genetic counselor prepared a pedigree for the Richardson family. Use the information you have about Emmet and his family to draw a pedigree like the one the genetic counselor prepared. Review the symbols in Figure 2 on page 60 to help you draw an authentic pedigree.

If Emmet's parents decided to have more children, what could you predict about their genotypes and phenotypes with regard to the sickle-cell gene? What can you predict about the genotypes and phenotypes of Emmet's children if he marries a woman with the same hemoglobin genes as his sister who has the trait? What if he marries a woman with the same hemoglobin genes as his sister who does *not* have the trait?

Emmet thought about his genotype. Although he knew that sickle-cell trait is not a disorder, he wondered about the children he might want to have someday. You and Emmet can learn about prenatal detection of sickle-cell trait and sickle-cell disease in "What Are RFLPs?" on page 122. 🖾

GENES AND THE ENVIRONMENT

HEREDITY, ENVIRONMENT, OR BOTH? HOW DO WE FIND OUT?

One of the oldest debates in human history centers around the role of heredity in various human traits. For example, is intelligence genetically determined? Is it the result of a stimulating environment? Does it result from a combination of the two? What about athletic ability? musical talent? susceptibility to disease? life span?

How do we separate the effects of heredity—nature—from the effects of the environment—nurture? Let's explore two methods geneticists use to help sort out that complex question: twins and adoptees.

TWINS

"Minnesota Group Studying Identical Twins" was the title of a magazine article Tim found one day while waiting in the dentist's office. Tim was a twin himself. And even though his friends teased him by asking if he and his twin were identical, it was pretty evident they were not, because his twin was his sister Kate. Even so, Tim found the article interesting and he told Kate about it later that evening.

"They studied lots of things about these twins, Kate," Tim told her. "One of the more interesting things was the similarities that showed up in both twins, even if they were separated at birth by adoption."

"I'll bet you mean *particularly* if they had been separated at birth. That would mean the similarities the twins have probably are not related to the different homes where they were raised," Kate interjected.

Kate was okay, Tim thought, but she could be a little bossy at times. Anyway, Tim had to admit she had it right. Tim and Kate had a lot of similar tendencies themselves, like the way they spoke and their love for anchovy pizza, but he figured it was because they grew up together.

Kate continued, "You know, it would be interesting to know if the separated twins still ended up with the same kinds of friends or jobs."

"Well, that's one of the things they looked at," said Tim. "In some cases, the twins ended up playing the same musical instrument or liked the same books or even had the same hobbies or in-

terests. Some of those things might have happened by chance, but I'll bet some are part of their genetics."

Just then Tim and Kate's mom came in. She had been at her desk working on a report for work and heard their conversation. "You know, kids, when you were just two years old, you participated in a genetic study of twins. A doctor at the university wanted to study identical and nonidentical twins to see if there were any differences that might not be genetic. It sounded interesting and important, so I took you down to the clinic, where they drew your blood. I remember Kate was as good as could be, but Tim cried quite a bit—and mostly before they did the test! We got a report a few weeks later telling us there were no abnormalities and thanking us for participating.

"A young geneticist named Kris organized the study. She said that many important conditions, such as depression and diabetes, have a strong genetic component. They occur much more frequently in identical twins than in nonidentical twins. Twins are usually brought up in the same house, with the same food and environment. So, if some trait is more common in identical twins, it is likely because identical twins have identical genetic material. Twins who are like the two of you only have as many genes in common as regular brothers and sisters."

That was all quite interesting to Tim and Kate, and they reflected on their differences and similarities as both went off to do their homework. Tim had to struggle with math and Kate had her problems with English.

Think about the material you just read as you develop answers to several questions your teacher will pose.

ADOPTEES

Scientists have long debated whether genetics or lifestyle plays the greater role in determining how long we live, and the answer has been elusive. But in March 1988, a group of Danish scientists reported that both factors exert a strong influence on life expectancy.

University of Copenhagen geneticist Thorkild Sorensen and his colleagues studied adoptees from 960 families. Studies of people who grow up in adoptive homes are common in genetics research. By comparing adoptees with both their biological and adoptive parents, geneticists can investigate the genetic and environmental factors that contribute to various traits and diseases.

If, for example, researchers find that children brought up in adoptive homes die at a relatively early age of the same diseases or disorders as their natural parents, then genetics was probably the biggest factor in their deaths. If, on the other hand, adoptees die prematurely of the same causes as their

adoptive parents, one can conclude that the home environment—including diet, alcohol consumption, smoking, and exercise—played a greater role in their deaths.

Directions

Select two or three classmates and work together to analyze the information presented in Tables 1 and 2. The information is based on Sorensen's study conducted in Denmark; he published the data in 1988.

Subjects: Adoptees in 960 families. All born between 1924 and 1926. All adopted by people unrelated to them.

Question: What were the causes of premature death in these subjects, in their biological parents, and in their adoptive parents? (Premature death means death between the ages of sixteen and fifty-eight.)

Questions for Discussion

Discuss the following questions in your group and record your answers on a separate sheet of paper. Be prepared to share your answers and explanations with the class.

1. According to Sorensen's data in Table 1, does heredity or environment play a greater role in general increased risk for premature death?
2. According to Sorensen's data in Table 2, which specific cause of premature death has the most significant genetic component? Which is most influenced by the environment?
3. Provide a general statement about the role of heredity and environment in the diseases Sorensen analyzed.

TABLE 1 General risk of premature death for adoptee.	
	Risk of Premature Death
If a biological parent died before age 50	risk is almost doubled
If an adoptive parent died before age 50	no increased risk

TABLE 2 Risk of premature death for adoptee based on cause of death in biological and adoptive parents.		
If biological parent died from	Risk of premature death for adoptee	If adoptive parent died from
infectious disease ⟶	fivefold	slight increase ⟵ infectious disease
heart disease ⟶	fourfold	tripled ⟵ heart disease
stroke ⟶	fourfold	fourfold ⟵ stroke
cancer ⟶	slight increase	fivefold ⟵ cancer

ALZHEIMER DISEASE: IS IT INHERITED?

John, Carlos, and John's grandfather had just returned home from a Cubs-Dodgers game at Wrigley Field. Carlos had a great time. He especially liked it when the Cub fans in the bleachers had thrown back onto the field a home-run ball hit by the Dodger catcher. He and John were talking about the game when John mentioned that his grandmother had been a great Cub fan, also.

"What happened to your grandmother?" Carlos asked.

"She died about three years ago. She had Alzheimer disease," John replied.

"I've never hear of Alzheimer disease. What is it?" Carlos asked.

"Well, I'm not exactly sure," John said. "When I was little, I remember we visited my grandparents every weekend. My grandmother would bake the world's best brownies, and she was like a best friend, always ready to play a game. Then, when I was about six or seven, she seemed to change. She would forget things I knew we had talked about, and sometimes she seemed a little cranky. Her memory got worse and worse, and she stopped taking care of herself. The last couple of years I hardly saw her at all, and my grandfather couldn't take care of her by himself. She died in a nursing home when she was sixty-two."

In Carlos's biology class a few weeks later, the teacher, Mrs. Leysens, was talking about Huntington disease. The symptoms of Huntington disease sounded like Alzheimer disease, so Carlos asked if they were the same thing.

"No, they're not, Carlos. Huntington disease is inherited, as you know. I'm not really sure if Alzheimer disease is. Perhaps you could do your biology report on it and we could all learn something."

Carlos liked to use the library, but when he did his reports he would first try to find someone to talk to about the topic and find out something interesting to focus on. Mrs. Holler, his next door neighbor, was a nurse, so he tried her first.

"Well, you're in luck, Carlos," Mrs. Holler said. "We had a patient with Alzheimer disease on our floor just two weeks ago. I learned that there is now good evidence that some forms of Alzheimer disease are inherited. If I remember correctly, the inherited forms sometimes show up earlier than the ones that don't seem to be inherited."

With that start, Carlos headed to the library. He learned that Alzheimer disease is similar to what happens naturally to old people. They lose their memory and their personalities change, but Alzheimer disease occurs earlier and progresses more rapidly. Doctors have found changes in some of the brain cells in these people. Surprisingly, these changes are similar to those seen in the brains of people with Down syndrome, a chromosomal disorder caused by an extra chromosome 21 (see "Robert Vandenberg Wins Special Olympics" on page 53).

That similarity led some scientists who map genes to suggest that a gene for Alzheimer disease might be on chromosome 21. In 1987, a group led by Dr. James Gusella of Harvard University found linkage between markers for chromosome 21 and families with Alzheimer disease. In the next couple of years, however, scientists found other families that did not have the gene on chromosome 21. That suggests that there is more than one gene for Alzheimer disease. In addition, there were still many cases that had no family history at all, so those seemed to have a nongenetic cause.

Once Carlos had finished giving his report in class, Mary Preece raised her hand. "That was interesting, Carlos. My great-grandfather has Alzheimer disease, also, but he's seventy-eight and no one else in our family has ever had it. I was wondering how you can tell if something is genetic. It must be hard to figure out for something like Alzheimer disease, where the people are so old when they get it."

"That's a good question, Mary," said Mrs. Leysens. "Let's divide the class into groups for the rest of the period and see if we can come up with some ideas on how to tell a genetic from a nongenetic condition."

You can do that in your class, also. Think about other causes of human disease, such as infections, exposures to toxic chemicals, or nutritional deficiencies.

An Episode of Depression

David and Peggy Johnson were in their parents' car on the way to their yearly Thanksgiving reunion at their grandparents' house in Kenmore, New York. It was about a ten-hour drive from their home in Springfield, so they had a lot of time to talk in the car. They had made this trip each of the last five Thanksgivings since they had moved from Kenmore. Despite the long drive, they really enjoyed seeing all the relatives with whom they had grown up. Mr. Johnson was the oldest of three boys and four girls, and all of them except his family and his youngest sister, Pam, still lived around Kenmore. Pam was a special favorite of David and Peggy's because she had babysat for them for many years. She was a junior at Carlton College now and had missed the last two reunions. But, she would be home this year.

"You both know that Pam was in the hospital for a month last year," said Mrs. Johnson. David had talked with his mom a lot about Pam last spring when they had heard she had gotten very depressed at college and had to be hospitalized. They had spoken with her on the phone several times since then, and Pam had explained that she felt fine now and was taking medication. Just before Pam had been hospitalized, David and Peggy had both received several letters from her. Although letters from Pam weren't unusual, these letters had been very long and didn't make any sense in some places. David had shown the last letter to his dad, who had phoned Pam and

sensed that something was wrong. Shortly after that, David's grandparents had visited Pam at college and they were the ones who helped get her into the hospital.

"Pam had an episode of depression," Mrs. Johnson continued. "Some people have a phase where they get very talkative and overly outgoing, like in the letter you got from Pam. Many doctors think some such episodes are caused by a genetically determined imbalance in the brain's chemistry. Fortunately, there are good medications available to treat the disorder, and Pam is doing just as well at school now as she was before. She told me last week that the doctors are taking her off the medication soon."

"If it's genetic, does that mean other people in our family also have it?" Peggy interjected.

"Well, they might," said Mr. Johnson. "My dad's sister Esther had some similar episodes when she was younger, but it wasn't really identified in her. She struggled for many years with her job and finally was unable to work. I always thought it was a shame they didn't have good treatment in those days, because Esther was bright. And you don't remember it, but your Uncle Paul also had an episode of depression when he was about twenty-three. He was on medication for about a year."

It was hard to think of Uncle Paul as sick in any way. He was the family joker who could make anyone laugh, and he had enough energy for three people. He had built a giant playground in his

backyard for his two little girls that was better than anything the schools had.

The trip was soon over, and when Peggy and David got out of the car Pam ran out to meet them. They thought her first words would be the standard, "You've grown so much," but Pam surprised them and said, "I've really missed you guys. You're the only ones who will play 'Kings in the Corners' with me!" During the next three days, they played a lot of cards and Pam told them all about school, running track, and her boyfriend Matt. To Peggy and David, she was the same old Pam, and it made them feel good that there were some genetic problems you could do something about.

On the way back home, their dad explained a bit more about manic-depression. In the manic stage, people get very talkative; in the depression stage, they feel very down on themselves and life in general. "Even though we all have times like that, in some people, those periods get serious enough to prevent them from functioning normally. That's when you need to get help. Sometimes it takes a friend or relative to see what's going on and get the person into treatment. As you saw, there are very good treatments and most people can quickly return to their normal lifestyles.

"Doctors have known for a long time that there is a genetic component, because the condition runs in families and is more common in identical twins than in regular brothers and sisters. In 1987, researchers identified a genetic marker in some families where the depression is inherited as a dominant trait. The discovery of the marker raises hopes that we can identify the exact abnormality. Most people develop the condition by their twenties or thirties, just as happened with Paul and Pam. It's interesting that my dad has never had an episode like Paul or Pam's. Doctors think that in addition to inheriting a gene for depression, something in your environment also must trigger the actual depression. Unfortunately, they have no idea what that might be."

David remembered that when they moved five years ago he had been very upset and depressed, but his mom reassured him that was a normal response for a nine-year-old leaving all his friends; it didn't mean he had manic-depression. Peggy said she was depressed spending twenty hours in the car in four days, and they all laughed. It was good to know it was okay to be depressed sometimes. It was even better to know that there are effective treatments for people like Pam.

INTELLIGENCE, I.Q., AND GENETICS

Whatever intelligence is, we value it highly in our culture. Therefore, information about high or low intelligence affects how we treat people in our society. What is intelligence? Here is one example:

Some people walk into a situation in which others are struggling, appraise it, and select an effective course of action. If they do such a thing only once, we may say they are lucky. If they can do it only in certain situations, we may say they have a special knack or talent. But what if they do it many times, in different situations for which their prior knowledge is no greater than yours or mine? If that happens, we say they are intelligent.

That seems to be the core, common sense meaning of intelligence. We can paraphrase it as general problem-solving ability, if we recognize that the problem comprises a wide range of situations, and that we infer abilities only from words or deeds. Others consider intelligence to be the ability to *learn* problem-solving skills.

There is no reliable test for such all-around problem-solving ability, if such an ability exists. Most people are good at solving certain kinds of problems and not good at solving others. Whatever people think of as intelligence, it is a complex combination of abilities.

WHAT IS I.Q.?

I.Q. is not the same as intelligence. We use it to measure intelligence, but there is no good way to do that. I.Q., or intelligence quotient, is a standardized score on certain tests. The tests—and there are several available—include items designed to measure a person's ability to recognize and to solve certain spatial, verbal, and mathematical problems.

There is cultural bias in the I.Q. tests. A test developed with one cultural group should not be used for another. For example, the tests in the United States, all of which were developed for English-speaking Whites, should not be used without modification for other groups, such as Spanish-speaking Americans.

To get an I.Q. number, we use a standardized score to allow easy comparison between scores. A person with an I.Q. score of one hundred got the median number of correct answers. That means that about fifty percent of the population on which the test was developed got higher scores, and about fifty percent got lower scores. A person with an I.Q. score of 115 scored higher than about eighty-four percent of the population on which the test was developed.

What influences a person's performance on an I.Q. test? We know that the score is partly a matter of the experiences a person has had. Is it also influenced by the genetic makeup of that person? We know that you don't inherit your I.Q. score in the same way you inherit your blood type. There is no single gene or pair of genes that controls whether you get a score of 95 or 105 on an I.Q. test. Of course, genes are involved in very basic ways, for they control body development, including the brain, from a fertilized egg to full maturity. They also are involved in maintaining all the chemical processes that keep you alive.

Why does it matter whether the score on an I.Q. test is determined more by genes or more by environment? In the best of all possible worlds, it would not matter at all. People would have unlimited educational opportunities, and everyone would learn to the best of his or her abilities. In our less-than-perfect world, however, the answer can influence how we treat individuals and groups of people. It also can influence policy decisions made by governments or schools. Answers to questions of how best to spend the nation's education budget rest on judgments about whether certain programs will "do any good."

In 1969, psychologist Arthur Jensen shocked a civil-rights-conscious America when he argued that average differences in I.Q. between Blacks and Whites in the United States are due in large part to inborn factors. Jensen's supporters maintained that special education programs like Head Start, therefore, are useless. In 1972, William Shockley, a physicist who won the Nobel Prize for his work on the development of the transistor, suggested that we encourage sterilization in people with I.Q. scores below one hundred so they don't pass on the genes responsible for the low I.Q. scores. Both men were strongly criticized for their methods of investigation and for their conclusions.

In 1994, Richard Herrnstein and Charles Murray published *The Bell Curve: Intelligence and Class Structure in American Life*. This widely read and controversial book echoed Jensen's assertion that racial differences in I.Q. scores are genetically based. In addition, the authors argued that a person's success in society is largely determined by his or her inherited intellectual ability.

GENES AND I.Q.

What is the evidence that genes have anything to do with I.Q. scores? Because I.Q. does not show simple dominant or recessive patterns, we know it is not controlled by one pair of genes.

Complex characteristics, like height, weight, or the ability to take an I.Q. test, are heavily influenced by the environment, as well as by many genes. We have developed a measure called heritability to estimate the amount of influence genes have on those kinds of traits. Heritability is the percentage of the variation in a population that is not due to environmental factors. If all the variation in I.Q. in a population was due to genetic factors, then the heritability would be 1.0, or one hundred percent, for the trait in question. If all the variation in a population was due to environmental factors, the heritability would be 0.0, or zero percent.

Research on twins has provided one way to estimate the heritability of I.Q. test-taking skills. Identical twins (monozygotic twins derived from a single fertilized egg) are genetically identical. Their I.Q. scores also are more similar than are the scores for individuals who are less closely related. Heritability measures calculated from twin studies vary, but may be as high as 0.8. That means that only twenty percent of the variation is assigned to environmental factors.

A study in 1964 found that when identical twins were raised in different families the average difference in I.Q. scores was eight points. When identical twins were raised in the same family, the average difference was six points. That suggests that the difference in environments has some effect, however slight.

New studies estimate heritability from data on extended families, including identical twins and their children. Here, researchers estimate the heritability of I.Q. scores is about fifty percent, which indicates the equal influence of genetic and environmental factors.

PROBLEMS OF DETERMINING HERITABILITY

There are two major problems when one tries to estimate genetic influence with measures of heritability using identical twins. First, it is difficult to separate genetic from environmental factors. Identical twins have almost identical environments if they grow up together. Even when they are separated and live in different homes, their environments probably are quite similar. Adoption agencies often try to place children with foster or adoptive parents who meet certain predetermined standards of age, income, educational background, and so on. That may account for some of the similarity among the I.Q. scores of identical twins. The extended-family technique suffers the same problem, because identical twins are likely to provide their children with similar environments. That makes it difficult to completely separate all of the genetic and environmental factors.

A second problem is that heritability measures cannot be applied to any population that is different from the one used to get the heritability estimates. If heritability of I.Q. scores is estimated at fifty percent in a population of White U.S. citizens with better-than-average incomes, we cannot apply that heritability estimate to other populations of humans, such as Asian Americans, White U.S. citizens with lower-than-average incomes, or the entire human population.

To understand that second problem better, consider height as an example. Suppose we wanted to estimate the heritability of height from a population of well-fed U.S. citizens. The heritability estimate would be high, because the environment would be similar for all individuals. Suppose, however, we included both well-fed and undernourished people in our sample population. In that sample, the measure of heritability would be much lower, because we included environmental differences.

In the same way, if heritability of I.Q. were measured for a population in which all individuals have similar environments, the estimate of heritability would be higher than if the sample included people living in a wide range of environments. That may explain why samples that include only White individuals with above-average incomes show high heritability, and studies using samples that include individuals from lower-income families show lower estimates of heritability of I.Q.

It also follows that differences in average I.Q. between populations cannot be attributed to genetic differences between the two populations. Blacks in the United States, on the average, score somewhat lower on I.Q. tests than do Whites. But if the tests had been developed for Blacks, Whites probably would score lower. In

summary, there is no evidence that any ethnic group is, on the average, any more intelligent than another ethnic group.

What does the available evidence suggest about genetic variability of intelligence in the normal range? In their textbook on human genetics, published in 1997, Vogel and Motulsky reply: "The answer is short: very little." They point out that some experts have reviewed the evidence and have suggested that genetic variability does not affect I.Q. performance at all. Other authors think the evidence indicates that heritability may actually be as high as the 0.8 value reported earlier. Vogel and Motulsky conclude: "Most scientists, if asked for an educated guess, will probably settle on values somewhere in between, more because they dislike extreme points of view than because of a strong conviction in favor of any positive evidence."*

GENETICS OF HEART DISEASE

After reading a magazine article on how lifestyle can affect one's risk for heart disease, Ben Fix's parents went on a health kick. Ben's mom had always played tennis, and now she began taking walks with two of her friends every evening. His dad started playing tennis twice a week, after not playing for almost ten years. His parents said those were the best ways they could exercise, because they both had jobs that kept them at a desk most of the day. The only hard part of their interest in health was that

both of them were dieting and trying to cut down on their fat intake, which meant that Ben's diet had changed, too. Luckily, they were sensible about not trying to do too much too soon, so the meals at home had only changed a little. After a month of drinking skim milk with two percent fat, Ben now drank skim milk with only one percent fat. Before, his family had always drank whole milk, which has four percent fat.

One afternoon, when Ben returned from band practice, his mom was already home and on

the phone. She seemed upset, and when she got off the phone she explained, "Your Uncle Jack has just had a heart attack; he's in the hospital intensive care unit. We're going to have Mike stay with us for a few days."

Mike was Ben's nine-year-old cousin, the son of Jack and Betty, who was Mrs. Fix's sister. It sounded as though Jack was okay

---

*Vogel, F. and A.G. Motulsky. 1997. *Human Genetics: Problems and Approaches*. New York: Springer-Verlag.

for now, Mrs. Fix continued, but she was going to the hospital and wanted Ben to stay home and tell his dad. She would bring Mike back with her.

A few weeks later, Jack was back at work and doing well. The family had discussed heart problems at the dinner table during the last few weeks, and Ben had learned that Jack's father had died from a heart attack at age fifty. Because Jack was only forty-six, it seemed to Ben that heart problems were happening pretty early. He had always been a worrier, and since he had learned that heart problems can run in families, he asked his parents what his chances were for a heart attack.

"That's a good question, Ben," said Mr. Fix. "Let's see if you can figure it out. Why don't you diagram the different people in our family and put down their ages and their health status. Then, we'll look it over and see if it tells us anything."

After dinner, Ben did as his dad suggested and came up with the pedigree shown in Figure 1. You will discuss this pedigree in class, so study it now and think about who in the pedigree might have a risk for heart problems of the kind that Jack and his dad, Nick, had.

Ben's dad, Steve, had read a lot about diet and genetic factors involved in heart attacks. He told the family that genetic factors have the biggest influence on your risk for developing early heart problems. He also said that scientists and doctors are improving their ability to identify those people early in life who are at risk for these problems. Men with these genetic risks often have heart attacks in their forties. Women at risk often have heart attacks while in their fifties. If identified early, people at risk can adopt a more healthy lifestyle, even as children, and perhaps prevent much of the damage that occurs over time. Ben thought it would have been nice to have gotten used to drinking skim milk when he was two, so he would not have to change now.

Mr. Fix said that in 1986 Drs. Michael Brown and Joseph Goldstein were awarded the Nobel Prize for having figured out the genetics and molecular biology of elevated cholesterol levels. That kind of work is leading to the possibility of identifying individuals at a high genetic risk who are most in need of having an appropriate diet and exercise program. Of course, the important point is that genetics *and* environment contribute to early cases of heart disease. You can't change the genes you inherited, but you can change your environment. For example, health experts have long recognized that the high-fat diets people in Western countries tend to eat present significant risks.

"This is just another example of how our genetic background and our environment interact with one another," said Mr. Fix. "You have to pay attention to both. But, I guess we're lucky to live at a time when scientists are figuring these things out. Besides, I have to admit I feel a lot better since I lost ten pounds and my backhand has improved. I might even win some games from your mother soon."

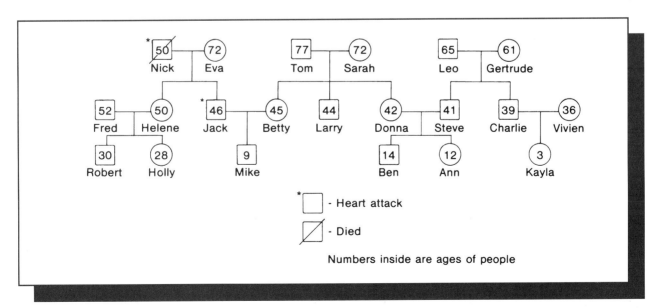

FIGURE 1 ▪ Ben's pedigree of the Fix family.

FEATURES

PRENATAL DIAGNOSIS

Many articles and activities in this magazine mention prenatal diagnosis. That alone justifies a feature on the subject. But, we have a more important reason than that. Today, any woman or couple planning to have a child—regardless of parental age, family history, or ethnic origins—will confront the possibility of becoming involved with one or more of the procedures that we describe in this article. Some will consider the procedures only to reject them immediately on moral and ethical grounds, but consider them they must. In fact, any physician who looks after pregnant women would find himself or herself in a very tenuous legal situation if he or she failed to discuss prenatal diagnosis with each pregnant woman or couple planning a pregnancy. It is worrisome to think that many couples are making crucial decisions about prenatal diagnosis without the understanding of the principles and procedures that you will have after working through this article.

There are those who oppose prenatal diagnosis of genetic and other disorders because they be-lieve that the procedures lead women to abortion. Women or couples who would not consider terminating their pregnancy for any reason ought not to feel pressured in any way to become involved in any prenatal diagnostic procedure. These are very personal decisions that each person or couple must make.

But, there is another way to look at prenatal diagnosis. Consider, for example, a couple whose first child is diagnosed at one year of age as having Tay-Sachs disease (TSD), a progressive, degenerative disease of the brain that inevitably leads to death in childhood. It is inherited as an autosomal recessive trait. Before the availability of the prenatal diagnostic test called amniocentesis (see Procedures section), few if any parents chose to take the chance of having more children with such a horrible, and as yet, untreatable disease. Today, we can detect TSD prenatally, and many parents make the opposite choice, taking comfort in the knowledge that seventy-five percent of the time the fetus and subsequent child will not have TSD. When the test reveals an affected fetus, couples may or may not decide to terminate the pregnancy—that choice is always the couple's. Few, however, undergo amniocentesis unless their minds are pretty much made up. The procedure itself increases the risk of miscarriage, and it would not be appropriate to put the pregnancy at risk if one did not intend to act on the test's results.

Dr. Barton Childs makes a pertinent point in the article "Why Study Human Genetics?" on page 4. He points out that close to fifty percent of all human conceptions are lost as early miscarriages (within the first three months of a pregnancy). The majority of those lost embryos and fetuses are abnormal; that is, they have multiple anatomical defects, lethal single-gene mutations, chromosome abnormalities, or other problems. Prenatal diagnosis simply detects some of the abnormalities that slip through nature's screening processes.

Curiously, the issue of research has raised a concern about prenatal diagnosis. Some people think that so much effort is going into identifying and characterizing more and more genes to allow more accurate diagnosis prenatally that less effort will be devoted to developing methods of treatment. That may be happening to some extent, but such major advances have been made in the treatment of genetic diseases that the concern is unlikely to be significant. We discuss many new approaches to treatment in this magazine; the following are two more exciting examples.

Fetal surgery, although still in its infancy (no pun intended), has already been successful. There is a birth defect called bladder neck obstruction, seen mostly in males, that is due to obstruction of the flow of urine from the kidneys and bladder to the urethra and out into the amniotic fluid space. The back pressure causes dilation of the whole urinary system and eventually destruction of the kidneys, often before the affected infant is born. If the infant is born alive, kidney failure usually occurs quickly and the baby dies.

The urinary tract dilation is easy to detect by fetal ultrasound (see Procedures section) and the fetal surgery consists of inserting a tube and one-way valve system, with ultrasound guidance, from the fetal bladder to the amniotic fluid, thus allowing the urine to drain out of the fetus and relieve the back pressure. The one-way valve prevents amniotic fluid from flowing into the fetus. Several pregnancies have been managed this way with preservation of kidney function. After birth, physicians operate to open the obstruction and the infant usually does well. That is just one of several approaches to prenatal surgery currently under intensive research.

One of the most intriguing advances in the prevention of birth defects is associated with spina bifida and related defects of the brain, spinal cord, and spinal nerves. The cause is unquestionably multifactorial and we have identified a number of environmental factors, including an inadequate intake by the mother of one of the B vitamins, folic acid. Several clinical trials, initially aimed at reducing the chance of a second affected child after a couple has had one with spina bifida, left no doubt that folic acid supplements taken during the first three months of pregnancy significantly lower the risk of recurrence. Later studies showed that if women who had *not* had a previous child with spina bifida were on similar folic acid supplements, the chance of having a first baby with one of those defects was significantly reduced.

A problem here is the embryology. The neural tube, which is the embryonic structure from which the brain, spinal cord and spinal nerves develop, completes its development from a groove running along the length of the embryo to a tube at about six weeks of gestation. That is just about the time that most pregnant women realize that they might be pregnant (they have missed one menstrual period). Thus, six weeks or later would be too late to start any intervention that might prevent problems with development of the neural tube. To have any preventive effects, women must begin taking the folic acid before they become pregnant; that was done for all of the women in the studies just mentioned. As a result, physicians now recommend that all women of child-bearing age who are considering having a child start on folic acid supplements before they stop whatever contraceptive technique they or their partner are using. At present, some evidence suggests that a multivitamin pill that includes folic acid may decrease the incidence of other birth defects, as well as those in the spina bifida group. It is too early to be sure of that or to know which other defects might be involved.

In the near future, we should see a measurable decline in the incidence of spina bifida and its related defects from its current occurrence of 1 in 700. The savings in health-care dollars (these children usually require extensive surgery and physiotherapy), although important, will fade to insignificance when compared to the avoided pain and suffering for the no-longer-affected children and their families.

THE PROCEDURES

Maternal serum screening (MSS). Although this is the most recent addition to the prenatal diagnostic techniques, we list it first because it is the only one that is being offered to *all* pregnant women. Its history goes back to the early days of amniocentesis, the 1970s, when researchers found that unique proteins produced by the fetus and the developing placenta leak out into the amniotic fluid. From the amniotic fluid, some of those proteins find their way into the mother's blood. We have carefully established the usual or normal levels expected at each week of pregnancy. Among the proteins are alpha-fetoprotein (AFP), produced mainly by the fetal liver, and human gonadotropic hormone (HGH), a product of the developing placenta. Those fetal proteins are unique only in that they are fetal and placental in origin and are not found normally in the blood of a woman who is not pregnant.

The first association between an abnormal AFP level and a birth defect was for a group of conditions known as spina bifida. Those are common birth defects, which occur in about 1 in 700 live-born babies. They also cause a large number of miscarriages and still-births (babies that are born dead). The problem occurs in the formation of the brain and spinal cord, and often results in an open defect that exposes parts of the spinal cord or the brain itself. Instead of the usual bones and skin, only a thin membrane separates the brain and/or parts of the spinal cord from the surrounding amniotic fluid. AFP simply leaks out in greater amounts than would leak through normal skin. If the fetus survives the pregnancy with a severe defect in the brain (the medical term is anencephaly, literally, absence of the brain in Greek), it will die shortly after birth. Live-born infants with defects in the spine and spinal cord often survive with symptoms that vary from no handicap to paralysis from the waist down with loss of control of bladder and bowel, the severity of which depends on the location of the defect and its size. Measurement of amniotic fluid AFP became a routine part of amniocentesis done for any reason, and as a result, we detected many cases of spina bifida prenatally.

Measurement of AFP in maternal serum came later, but before we look at that issue, let's make sure we understand the difference between a screening test and a diagnostic test. Stop for a moment;

can you define those two types of tests?

We use a screening test with relatively large groups of people (in this situation, the entire population of pregnant women) to identify those who are at high risk of something but otherwise show no detectable signs of it. Inevitably, there will be false negatives and false positives. Think about screening for gold in a stream— among the gold nuggets that stay in the pan (the true positives), we will find the larger stones and bits of dirt (the false positives); the smaller pieces of gold will pass through the screen and be lost (the false negatives).

On the other hand, individuals, not populations, benefit from diagnostic tests. Usually, the tests are highly accurate; that is, there are few false positives or false negatives. Chromosome analysis or direct molecular detection of a mutant gene are good examples.

The first attempts at MSS used only AFP, and when the ranges of AFP levels were defined, physicians picked up close to seventy percent of the cases of spina bifida and nearly all the cases of anencephaly in the screen. The diagnostic test is ultrasound, which we describe under ultrasonography.

As specialists conducted the serum AFP studies on more and more pregnant women, they observed that when the mother was carrying a fetus with Down syndrome and some of the other much less common chromosome abnormalities (trisomy 18, for example), the serum AFP tended to be low. Adding maternal serum HGH levels and other fetal and placental proteins to the equation has increased the true positives for both Down syndrome and spina bifida, while it decreased the false

positives and false negatives. For example, most MSS programs use a low serum cut-off point that selects six percent to eight percent of the screened women as positive and "eligible" to go on to amniocentesis, the diagnostic procedure, if they wish. Most do have the diagnostic test, because they decided in advance to proceed through the entire process if found to be at high risk by the screen. In that six percent to eight percent of screened positive pregnancies, we find sixty percent to seventy percent of the fetuses with Down syndrome. Nevertheless, even among that six percent to eight percent of positive pregnancies on the screen, the majority will not have an abnormal fetus when the chromosomes are tested after amniocentesis. (The probability works out to about 1 in 8–10 true positives.)

At present, there are several state and provincially supported MSS programs available to all pregnant women in the U.S. and Canada who want to have this type of prenatal testing. In most areas without funded programs, the test is available at a reasonably low cost. Brochures in many languages have been published to help ensure that no woman embarks on this, or any other prenatal detection program, without understanding the potential implications.

The MSS testing is done at fifteen to sixteen weeks when we are best able to distinguish between normal and abnormal levels of the fetal/placental proteins. All rise and fall as the pregnancy proceeds.

MSS is having an impact on what used to be the most common indication for prenatal diagnosis, advanced maternal age. As shown in Table 1, the frequency of Down syndrome increases fairly gradually with maternal age until the middle to late thirties, when it begins to rise quite rapidly. Most geneticists felt that amniocentesis should be offered as an option to women age thirty-five or older, although thirty-five was clearly recognized as arbitrary. Close to fifty percent of pregnant women over thirty-five, on the average, were having amniocentesis in the U.S. and Canada for that reason. More recently, however, with the availability of MSS, many women in their thirties are deciding to have the screening test first. If the result drops their risk, they usually decline amniocentesis, and if the serum test raises their risk, they go on to that diagnostic procedure. The reasoning behind this is that a blood test is risk free; amniocentesis and chorionic villus sampling increase the chance of miscarriage as a result of the procedure.

Thus, MSS programs have actually reduced the number of women in their thirties who have amniocentesis. In fact, that decrease is the reason for the small increase in the total number of amniocenteses done for all pregnant women regardless of age in spite of the availability of MSS for all pregnant women. Most geneticists and obstetricians continue to recommend amniocentesis for women who are in their late thirties and forties because of the higher risks and concerns that MSS results will be misleading (too many false negatives).

Amniocentesis. This procedure involves withdrawing from the uterus a small amount of the fluid that surrounds the fetus (see Figure 1). Amniocentesis usually is done between the fourteenth and sixteenth weeks of pregnancy (counted as weeks after the first day of the last menstrual period). The obstetrician inserts a needle through the woman's skin and uterine wall and into the amniotic sac. The procedure is done in a clinic, and it does not require hospitalization. The small amount of fluid that is withdrawn during the procedure is replaced naturally in about four hours. The procedure does have a small risk—about 1 in 200—of causing a miscarriage.

The fluid withdrawn from the amniotic sac contains living cells from the fetus. We can grow these cells in a cell culture (see "Studying Human Chromosomes," page 56). The culturing and subsequent tests can take three or four weeks. Lab technicians use the cultured fetal cells to determine the fetal karyotype or to look for certain biochemical or DNA abnormalities. In some cases, the amniotic fluid itself is used for the prenatal diagnosis, as described earlier.

Ultrasonography. Before the obstetrician begins amniocentesis, the patient's abdomen is scanned with ultrasound. High frequency vibrations are translated into an image of the fetus and the surrounding structures in much the same way as sonar equipment is used on ships to identify submerged objects. The ultrasound image appears on a television screen, and the viewers, including the parents, can actually see the fetus "swimming" in the amniotic fluid (see Figure 2). The image allows the physician to obtain an accurate estimate of the fetus's age by determining its size. The picture also shows the location of the placenta and the fetus, so the doctor can avoid injuring either during the amniocentesis. Physicians can identify twins in the image and often the sex of the fetus.

Ultrasound pictures are now clear enough to allow a physician to identify several fetal abnormalities, including enlarged (cystic) or absent kidneys, hydrocephalus (water on the brain), certain forms of dwarfism, spina bifida, and even some cases of

TABLE 1 Frequency of Down syndrome infants among births, by maternal age*.	
Maternal Age	Frequency
20	1/2000
30	1/885
35	1/365
38	1/176
40	1/109
43	1/53
45	1/32
48	1/16

*These frequencies have remained essentially the same since the publication of this study in 1978.
(Hook, E. G., and A. Lindsjo. 1978. "Down Syndrome in Live Births by Single Year Maternal Age Interval in a Swedish Study: Comparison with Results from a New York State Study." *American Journal of Human Genetics* 30:19.)

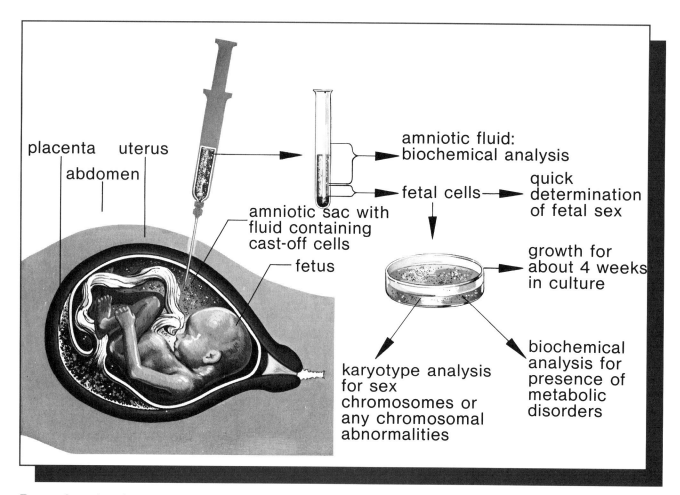

placenta   uterus
abdomen

amniotic fluid: biochemical analysis

amniotic sac with fluid containing cast-off cells

fetal cells → quick determination of fetal sex

fetus

growth for about 4 weeks in culture

karyotype analysis for sex chromosomes or any chromosomal abnormalities

biochemical analysis for presence of metabolic disorders

FIGURE 1 ▪ Amniocentesis.

congenital heart disease (defects in the structure of the heart that are present at birth). This technique also may permit detection of severe hereditary diseases of the muscles, brain, and spinal cord, where fetal movements may be diminished or absent.

Chorionic villus sampling (CVS). This newer procedure is done at ten to eleven weeks of pregnancy. Physicians insert instruments through the vagina and cervix or the abdominal wall and into the uterus (Figure 3). Small samples of fetal cells are taken from the chorionic villi. The villi are fingerlike projections that grow out from the developing embryo into the wall of the uterus;

they eventually become part of the placenta. The cells in the villi grow rapidly. Results of the tests are available in one to ten days.

The risk of spontaneous abortion after CVS is increased by about one percent, which is considerably higher than the risk for amniocentesis. Nonetheless, its use is increasing, because it provides an earlier diagnosis for many conditions (but not all), and it can be performed a few weeks earlier in the pregnancy than amniocentesis. An early diagnosis is especially helpful if termination of the pregnancy is an option. For many people, termination early in pregnancy is more acceptable than later. In addition, it is the proce-

dure of choice for disorders whose diagnosis requires DNA analysis. CVS provides more cells more quickly. Some DNA tests can be done only in one or two laboratories around the world, and the procedures can be difficult and time consuming.

Cordocentesis. With this technique, doctors insert an amniocentesis needle directly into a blood vessel (often in the umbilical cord) of the fetus. They withdraw blood for rapid chromosome studies or to check the fetus for anemia or other blood problems. Medication or blood transfusions also can be provided in a similar manner.

X-rays. X-rays have limited use in prenatal diagnosis. Because the

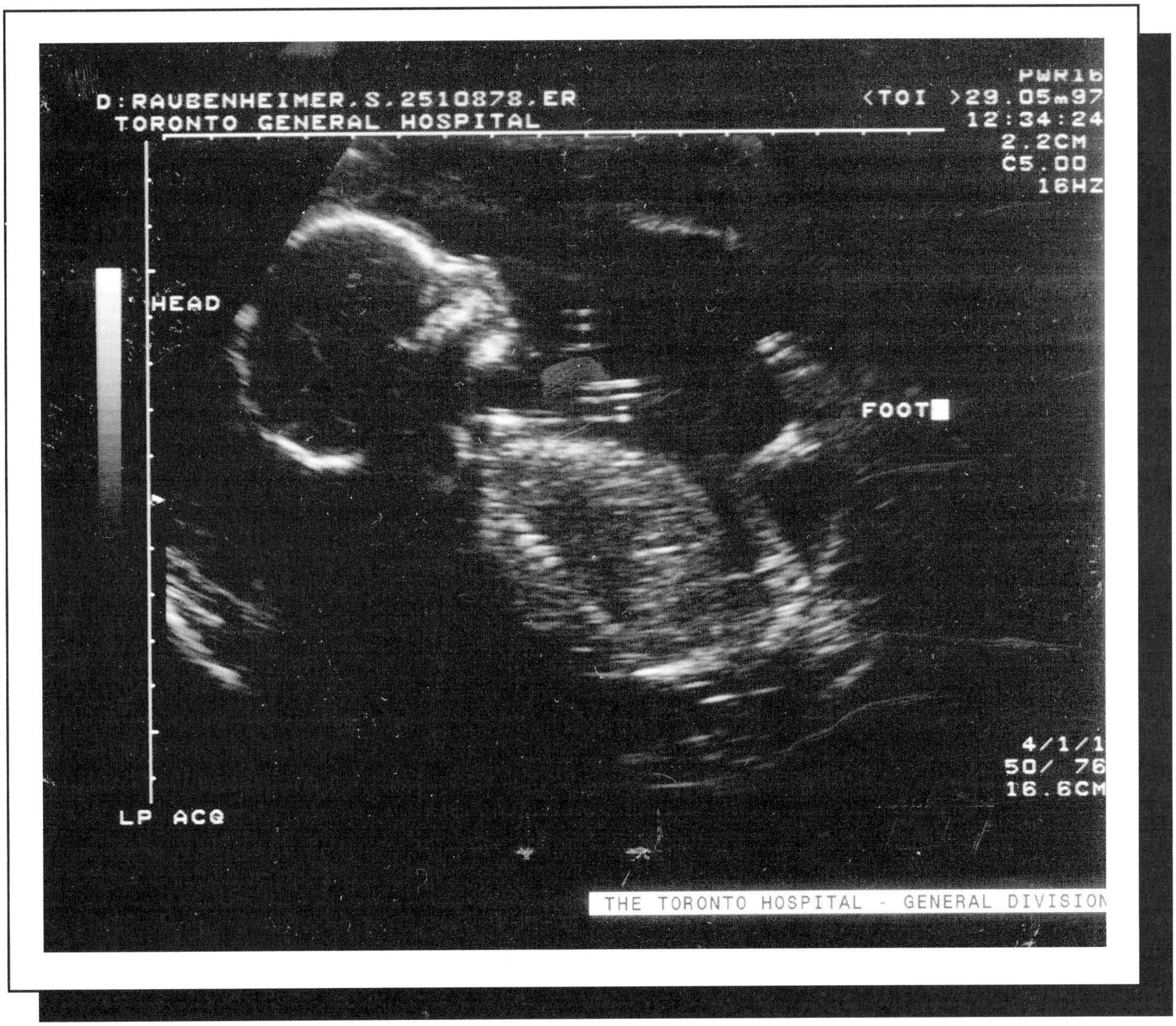

FIGURE 2 ▪ Ultrasonograph. (Courtesy of Dr. Ants Toi, Radiology Department, The Toronto Hospital, Toronto, Canada.)

fetal skeleton is only slightly mineralized between sixteen and twenty weeks of a pregnancy, it is difficult to see very much. In addition, X-rays potentially are a hazard to the fetus. The risk of a miscarriage is quite high—about five percent.

Fetoscopy. Fetoscopes, which use fiber optics, allow the doctor to visualize directly small areas of the fetus and to obtain skin or blood samples directly. Fetoscopy is now very limited, because the risk of miscarriage is high and because advances in ultrasound and cordocentesis can provide most of the same information.

Chromosomal analysis. Doctors can examine the chromosomes of cultured cells for abnormalities in number or structure. Although fetuses with Down syndrome (trisomy 21) represent the most frequently encountered chromosomal problem, several other chromosomal abnormalities, rearrangements, and deletions lead to severe mental and physical defects and to early death of the infant.

A technique called fluorescence *in situ* hybridization (FISH), developed in the early 1990s, allows detection of some chromosomal abnormalities in a matter of days rather than weeks. FISH does not require the use of chromosomes derived from cultured cells because the analysis is not based on a standard karyotype. Instead, FISH uses fluorescent DNA probes that bind to selected regions of specific chromosomes. The probes can be used on cells that are not undergoing mitosis. The fluorescent probes show up under ultraviolet light. Figure 4 shows cells that have three X chromosomes. In this case, the fluorescent probe is attached to DNA that is near the centromere of the X chromosome.

Biochemical analysis. About one hundred inborn errors of metabolism have been diagnosed prenatally through biochemical analysis of cultured cells or, in some cases, the fluid itself. Some disorders, such as Tay-Sachs disease and galactosemia, are diagnosable through biochemical analysis. Still another analysis can detect levels of alphafetoprotein (AFP) in the amniotic fluid, as described in the section on maternal serum screening.

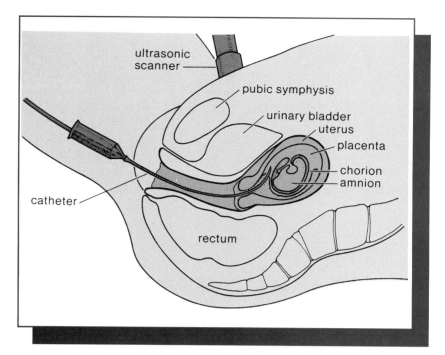

FIGURE 3 ▪ Chorionic villus sampling (CVS).

DNA ANALYSIS

Technicians can isolate DNA from chorionic villi cells or amniotic-fluid cells and analyze it for genetic disorders in several ways. For disorders such as cystic fibrosis, sickle-cell disease, or Huntington disease, where we know the mutation that causes the disorder, we can study the DNA directly for the presence or absence of the mutation (see "What Are RFLPs?" on page 122). For diseases where the mutation is not yet known but a linkage marker has been identified, the DNA of the fetus and other important family members can be tested for linkage; that will provide an accurate assessment of whether the fetus has inherited the affected gene. As more genes are identified and cloned, DNA testing is replacing biochemical

testing. DNA testing can be limited in some cases because it requires the study of particular family members who may not be willing to cooperate or who may have died.

WHO SHOULD HAVE PRENATAL DIAGNOSIS?

The procedures described in this article allow all couples at an increased risk of having children with severe genetic disorders to consider taking steps to reduce that risk. Almost always, the chance of having a normal child is far higher than the risk of the problem. In most cases, therefore, the test results are normal, and the parents are reassured. When a positive test indicates that a fetal abnormality is likely, physicians and geneticists present all possible options and discuss them with the couple. Some parents decide to

use the remaining months of the pregnancy to prepare themselves emotionally to deal with the problem their baby will almost certainly have.

With increasing frequency, the information we obtain through prenatal testing is used to ensure the best possible obstetric and pediatric care for the infant and the mother. Suppose, for example, that a woman in a small town is found to be carrying a fetus that has severe spina bifida, and she and her husband decide to continue with the pregnancy. When the pregnancy is near its term, the family doctor would transfer the mother to a large medical center where a team of experts would be assembled to handle the delivery. A neurosurgeon and a pediatrician would be in the delivery room as well, to supervise the immediate care of the baby and to decide whether emergency surgery is needed.

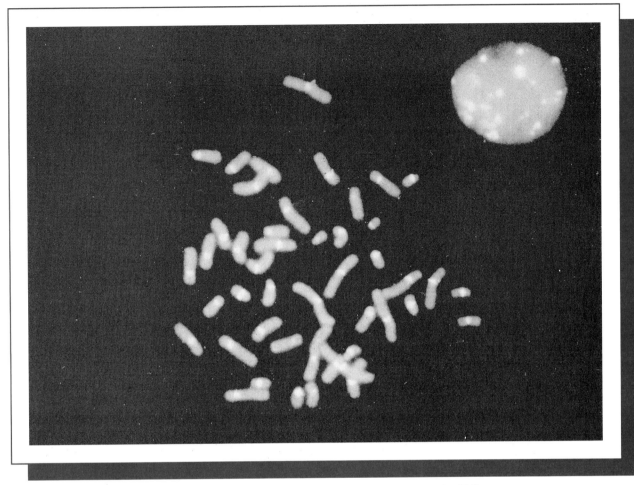

Figure 4 ▪ Chromosomes analyzed by fluorescence *in situ* hybridization (FISH).

In some rare instances, it is possible to treat the fetus before birth. For example, health-care specialists are now designing special diets for the mother that will provide the best possible chemical environment for certain enzyme-deficient fetuses. In addition, some surgical interventions performed prior to birth have been successful in a small number of cases as described in the introduction to this activity.

After much thought—and perhaps discussion with their physicians, geneticists, relatives, and others—some couples decide that the fetal defect is sufficiently serious to justify an abortion. Keep in mind that negative test results do not guarantee that the fetus is "normal." The majority of birth defects, even today, are not detectable by any of the available procedures.

As with many new biomedical techniques, the implications of amniocentesis and CVS have been far-reaching and controversial. Much of the moral controversy concerns abortion. Women found to be carrying a fetus affected by a chromosomal, biochemical, or neural-tube disorder can elect to terminate the pregnancy. The Supreme Court of the United States—*Roe v. Wade, Doe v. Bolton,* 1973—determined that a pregnant woman may legally elect abortion during the first and second trimesters of pregnancy, yet the abortion controversy continues. In 1989, the court ruled—in *Webster v. Reproductive Health Services*—that individual states may limit access to abortion. The debates continue in the legislatures of several states and will almost certainly reach the Supreme Court again.

Much of the controversy surrounding abortion—at least as it relates to genetics—results from the conflict between new technologies and traditional values. Prenatal diagnosis, because it affects decisions about reproduction, family life, and the quality of life, is particularly controversial in that regard.

MORAL CONTROVERSIES

Consider the following issues related to prenatal diagnosis:

1. Do biology teachers have a responsibility to teach their students about issues in human genetics that might affect the students' future decisions about marriage and having children?

2. Are the present criteria for eligibility for prenatal diagnosis reasonable?

3. When, if ever, should prenatal diagnostic testing *not* be performed—even if the woman meets the standard eligibility criteria?

4. Is the patient's physician morally required to inform a patient of the availability of prenatal diagnostic procedures? Should a physician be legally required to provide such information?

5. What should the physician do if he or she is morally opposed to abortion or to the possible use of the information gained from prenatal tests that might lead to a consideration of abortion?

6. Should abortion be performed if, as a result of prenatal diagnosis in the second trimester of pregnancy (twelve to twenty-four weeks of gestation), doctors find that the fetus will become severely mentally retarded or physically disabled?

7. Prenatal testing for genetic disorders also reveals the sex of the fetus (and, as in most cases, the fetus is known to be unaffected). Is a decision on the part of the parents to abort the fetus on the basis of its sex alone morally acceptable?

8. Should the patient's physician withhold information about the sex of the fetus from the parents in cases where the results of the prenatal diagnostic tests are negative?

9. Should we establish laws to compel prenatal diagnosis for all pregnant women who are at high risk for having a child with a serious physical or mental disorder, if that disorder is detectable prenatally?

10. Should a couple have the option to abort a pregnancy when prenatal diagnosis—carried out because of an assumed high risk for a serious disorder—reveals instead a completely different, less serious condition?

11. Should a couple abort a pregnancy when twins are present and one of the pair is found to have Down syndrome or some other serious disorder? Would it be acceptable to abort or otherwise eliminate only the affected fetus? (That is possible, and it has been done. The risk of losing both twins as a result of the procedure is high.)

12. Should every pregnancy be followed by ultrasound examinations of the fetus? Many obstetricians do at least two ultrasounds routinely to assess growth and to detect multiple fetuses (twins, triplets, etc.). But what if ultrasound reveals an unexpected fetal abnormality and the couple is opposed to abortion and did not wish to have that information in advance?

GENETICS IN HISTORY: GREGOR MENDEL

The life and the accomplishments of Gregor Johann Mendel, the Augustinian monk whose experiments with garden peas and flowers form the basis of modern genetics, are often romanticized. He tends to be portrayed as a simple monk who almost stumbled onto the laws of heredity as he tended the monastery garden. With no intention to diminish the extraordinary contributions of this brilliant man, it is important to realize that few major advances occur "out of the blue." Most successful investigators are well educated, hard-working, and familiar with the related work of their predecessors. They are innovative individuals whose work builds on data obtained by others. Mendel was no exception.

Gregor Johann Mendel was born on 22 July 1822 in Heinzendorf, an agricultural district of present day Czechoslovakia. The son of a peasant, Mendel received much of his early education from the vicar of a neighboring village.

Gregor Johann Mendel. (Courtesy of Dr. Vitězslav Orel, Mendelianum, Moravian Museum, Brno, Czech Republic.)

Mendel's early education included horticulture, as well as elementary subjects. Mendel's teachers recognized his outstanding ability at an early age and advised his parents to send him to the secondary school in Leipnik. He was later sent to the Gymnasium (secondary school) in Troppau.

During Mendel's fifth year at Troppau, his father became seriously ill and was forced to retire. Mendel began private tutoring to pay for his education. Weakened by the strain of long hours, Mendel became very ill and was forced to leave the Gymnasium for several months. He later recovered, completed his studies, and enrolled at the University of Olmütz. Beset by illness again, Mendel left the university and spent a year recuperating at his parent's home. With the aid of his younger sister, who gave him a part of her dowry, Mendel was able to complete his education at Olmütz.

On the recommendation of his physics teacher, Mendel entered the Augustinian monastery. The decision to enter the monastery at Brno was a matter of necessity rather than choice. At the monastery, Mendel was able to continue his studies without the financial worries that had plagued him during the previous years. During his stay at the monastery, Mendel studied agriculture and sheep breeding, as well as philosophy and natural science. He was later placed in charge of the experimental gardens run by the monastery.

When Mendel completed his theological studies, he was assigned duties at a neighboring hospital where he ministered to the sick. Mendel was so overcome by the condition of the patients that he began to experience severe depression. The abbot transferred him to Znojmo, in Moravia, where he served as a substitute teacher.

Mendel was so successful and popular as a teacher, it was suggested that he take a university examination for teachers. Mendel took the examination, but failed it. On the advice of the physics professor, Mendel was sent to the University of Vienna.

After returning to Brno, Mendel taught at the Brno Technical High School. During this time, Mendel began his work on garden peas. In May of 1856, he attempted to take the university examination for teachers for the second time. Unable to handle the stress associated with the examination, Mendel broke down and did not complete the examination. He returned to Brno very ill. He served as a substitute in the Brno high school until 1868, when he was elected abbot of the monastery. He served in that capacity until his death in 1884.

At about the same time Mendel was beginning his plant hybridization experiments, he began a careful study of weather. He quickly achieved a reputation as an authority on meteorological phenomena. He published numerous meteorological reports, beginning in 1863. As early as 1870, he suggested that the collision of conflicting air masses produced tornadoes. Although Mendel's weather studies received less recognition than his plant hybridization experiments, they demonstrated the same research methods that characterized his plant experiments.

Mendel's plant experiments were unique for several reasons. First, he bred many plants with identical traits. He later made hundreds of crosses for each trait. Second, Mendel used the mathematics of probability to analyze his data and to arrive at hypotheses that would explain his results. Third, Mendel did not try to study every characteristic of an offspring at once. Instead, he limited his study to a single trait at a time.

During Mendel's time, scientists knew nothing about chromosomes or cell division. Using only the results of his breeding experiments, Mendel provided the first clear explanation of the nature of heredity.

Although Mendel published his results in 1865, it was not until 1900 that his paper was discovered by three scientists working independently on similar problems. All three recognized that the report of an almost unknown Augustinian monk preceded their reports by more than three decades. Mendel already had laid the foundation for the study of heredity.

Although Mendel lived at the same time as Charles Darwin, who developed the theory of evolution by natural selection, historians of science generally agree that Darwin was unaware of Mendel's work. That was unfortunate for Darwin, because Mendel had part of the answer to a problem Darwin could not solve: how to explain the origin and inheritance of the variations within species that provide the basis for natural selection. Mendel had discovered that those variations arise in and are transmitted by those biological factors we now call genes.

To learn more about Gregor Mendel, you can consult an excellent biography titled *Gregor Mendel: The First Geneticist*, by Vitězslav Orel, Oxford University Press, 1996. Dr. Orel, himself a geneticist, was for many years the director of the Mendel Museum in Brno, in what is now the Czech Republic.

BE AN ACTIVE PARTICIPANT

With your new understanding of genetics and of some of the genetic disorders that affect humans, you may want to learn more or do something to help someone. Here are some suggestions:

1. If you live near a university medical center or a major metropolitan hospital, you may find an active genetic services unit. Contact them and let them know of your interest in genetics. Try to state a specific area of interest. If you are too vague or general, busy professionals may not want to take the time to work with you. Visit the unit or department to find out what they do.

2. Check to see if there is any volunteer work related to genetics in your community. Your local hospital might need your help with health-care activities involving children or adults with genetic disorders. Nursing homes, day care centers, and nursery schools might have people with genetic disorders. They probably would be grateful for your help. Many groups need volunteers, and your experiences as a volunteer will teach you a great deal.

3. Get involved with one or more of the many genetic voluntary organizations. There is much you can do for them: clerical work, telephone work, active participation in fund-raising, and organizing fund-raising activities—walkathons, bikeathons, runathons. If you don't know where to start, contact one of these agencies. Tell them what you would like to do. They will direct you to an agency that will be glad to have your help.

GENERAL ORGANIZATIONS

National Society of Genetic Counselors, Inc.
233 Canterbury Drive
Wallingford, PA 19086-6617
Phone: (610) 872-7608
FAX: (610) 872-1192
E-mail: nsgc@aol.com
http://members.aol.com/ nsgcweb/nsgchome.htm

Devoted to the improvement and promotion of genetic counseling. NSGC can provide general information about genetic counseling and about training and careers in that field.

American Society of Human Genetics
9650 Rockville Pike
Bethesda, MD 20814
Phone: (301) 571-1825
FAX: (301) 530-7079
http://www.faseb.org/genetics

Devoted to research and education in all areas of human genetics. ASHG can provide information about careers in human genetics.

March of Dimes Birth Defects Foundation
1275 Mamaroneck Avenue
White Plains, NY 10526
Phone: (914) 428-7100
FAX: (914) 428-9366
http://www.modimes.org

Eleven thousand local units. Concerned with the prevention of all birth defects. Write for publications about how to help ensure the birth of a healthy baby.

National Human Genome
    Research Institute
Office of Communications
9000 Rockville Pike, Building 31,
    4B09
Bethesda, MD 20892
Phone: (301) 402-0911
http://www.aamc.org/
    research/adhocgp/nchgr.htm
    An institute of the National In-
stitutes of Health. Shares the
management of the Human Genome Project with the U.S. Department of Energy. Conducts
research and provides educational materials about advances
in genome research.

Alliance of Genetic Support
    Groups
35 Wisconsin Circle, Suite 440
Chevy Chase, MD 20815-7015
Phone: (301) 652-5553 or 1-800-
    336-GENE
FAX: (301) 654-0171
http://www.mdacc.tmc.edu
    Provides information about or-
ganizations that help individuals
and families affected by genetic
disorders. Also provides information about public-policy issues
related to genetics.

ORGANIZATIONS FOR SPECIFIC CONDITIONS

This list contains only a few of
the many organizations that exist to help individuals and families affected by specific genetic
disorders.

Cleft Palate Foundation
1218 Grandview Avenue
Pittsburgh, PA 15211
Phone: (412) 481-1376 or 1-800-
    242-5338
FAX: (412) 481-0847
http://www.cleft.com

Cooley's Anemia Foundation,
    Inc.
129-09 26th Avenue, Suite 203
Flushing, NY 11354-1131
Phone: (718) 321-2873 or 1-800-
    522-7222
FAX: (718) 321-3340
http://www.stepstn.com/nord/
    org

Cystic Fibrosis Foundation
6931 Arlington Road
Bethesda, MD 20814
Phone: (301) 951-4422 or 1-800-
    344-4823
FAX: (301) 951-6378
http://www.cff.org

National Down Syndrome
    Congress
1605 Chantilly Drive, Suite 250
Atlanta, GA 30324-3269
Phone: (404) 633-1555 or 1-800-
    232-6372
FAX: (404) 633-2817
http://www.carol.net

National Foundation for Jewish
    Genetic Diseases, Inc.
250 Park Avenue, Suite 1000
New York, NY 10017
Phone: (212) 371-1030
http://www.stepstn.com/nord/
    org

National Hemophilia
    Foundation
110 Greene Street, Room 303
New York, NY 10012
Phone: (212) 219-8180 ext. 3049
    or 1-800-424-2634
FAX: (212) 966-9247
http://www.ortge.ufl.edu

Huntington's Disease Society of
    America
140 West 22nd, 6th Floor
New York, NY 10011-2420
Phone: (212) 242-1968 or 1-800-
    345-4372
FAX: (212) 243-2443
http://www.aclin.org

Muscular Dystrophy Association
3300 East Sunrise Drive
Tucson, AZ 85718-3208
Phone: (602) 529-2000
FAX: (602) 529-5300
http://www.ortge.ufl.edu

National Neurofibromatosis
    Foundation, Inc.
95 Pine Street, 16th Floor
New York, NY 10005
Phone: (212) 344-NNFF or 1-800-
    323-7938
FAX: (212) 747-0004
http://www.nf.org

Children's PKU Network
10515 Vista Sorrento Parkway,
    #204
San Diego, CA 92121
Phone: (619) 587-9421
FAX: (619) 450-5034
http://
    www.familyvillage.wisc.edu

The Foundation Fighting
    Blindness
1401 Mount Royal Avenue, 4th
    Floor
Baltimore, MD 21217-4245
Phone: (410) 225-9400 or 1-800-
    683-5555
FAX (410) 225-3936
http://www.blindness.org

Sickle Cell Disease Association
    of America, Inc.
200 Corporate Pointe, #495
Culver City, CA 90230-7633
Phone: (310) 216-6363 or 1-800-
    421-8453
FAX: (310) 215-3722
http://www.stepstn.com/nord/
    org

Spina Bifida Association of
  America
4590 MacArthur Boulevard, NW
  #250
Washington, DC 20007-4226
Phone: (202) 944-3285 or 1-800-
  621-3145
FAX: (202) 944-3295
http://www.charitynet.org

National Tay-Sachs and Allied
  Diseases Association, Inc.
2001 Beacon Street, Room 204
Brookline, MA 02146
Phone: (617) 277-4463
FAX: (617) 277-0134
http://mcrcr4.med.nyu.edu

PEOPLE

GENETIC SCREENING: PREVENTION WITH PROBLEMS

Soliloquy on Screening*
*(With apologies to William
   Shakespeare)*

To screen or not to screen
That is the question!
Whether it is nobler to proceed
With a test for mutant genes

Only after the minds of all have
   been prepared
By proper education,
Or to begin to test, anon, because
It is the thing to do.
One should not ask
To test
Without informed consent!
Alas, in time
Ignorance and confusion
In the minds of parents and
   screenees
May cause pain, suffering and
   stigmatization
To those innocents who ask you
For the genes they are heir to.
And, may at some distant day
Defame those who screen.
For whether one should test a
   pound of flesh,
A single cell or a drop of blood

---

*Robert F. Murray, Jr., M.D., Howard
University, Washington, D.C.

It is that person tested who must
Live with and adjust to
The label "carrier"
And therein lies the rub!

The year is 2010. Linda, a new-born baby, is a few hours old. Samples of her blood, urine, skin, and placental tissue go into the hospital's computer system. Three red lights flicker on the screen. She is a carrier of three genetic disorders. That fact is placed on her Universal Health Card. The card will be deposited in Washington, D.C. Linda's card has all the genetic information known about her. Each newborn baby has a similar card.

Twenty-one years later Linda and her boyfriend Mark apply for a marriage license. At this time, a computer compares Mark and Linda's health cards. It turns out that Mark and Linda both carry a gene for the same serious defect—a recessive disorder. Any child they may have has a one-in-four chance of being mentally retarded.

Is this story a prediction of the future or just a fantasy? When Mark proposes to Linda, will she say, "Well, first we'll need our computer health-card clearance,

and then . . ."? How far-fetched does that sound? Well, it is not as far-out as you might think. Birth registries already exist in many areas of North America, the United Kingdom, and Europe. Information obtained from birth certificates and medical records is entered into a computer. If, for example, the frequency of certain birth defects begins to increase sharply in a particular geographic area, the computer signals the increase. Investigators can then determine whether a mutagen or teratogen recently introduced into the environment might be the cause.

On an experimental basis, a few genetics centers have devised a genetics record-linkage system. Again, we use computers, this time to link members of a family together. Family histories are stored in the computer's memory bank along with data on possible genetically determined conditions. One such system, for Huntington disease, is used to tie families together for the DNA studies reported in "Discovery of the Gene for Huntington Disease" on page 50. Access to the information in that databank is restricted and available only with the permission of the program participants.

Registries and the computer storage and the retrieval of genetic information are but two outgrowths of our increasing ability to screen for genetic disorders. In fact, genetic screening is becoming a routine medical procedure for many conditions. "The Screening of Benjamin Miller" (page 85) describes the first genetic screening program ever carried out on an entire population—in this case, nearly all newborn babies. The article also reveals how such screening programs began.

The National Academy of Sciences has made it clear that genetic screening is different from most other types of medical screening, such as that for infectious diseases. Most children have tuberculin tests for TB, and hearing and vision tests in school. Those are all nongenetic screening tests. (Keep in mind, of course, that genetic factors play a role in the susceptibility to infections and in the development of many kinds of hearing and visual problems.) Such tests benefit the individual who is screened. Genetic screening, on the other hand, can benefit both the individual screened when treatment is available, and the family, by providing information about the likelihood that the condition will occur again.

Sometimes, genetic screening can create as many problems as it solves. Newborn babies are routinely tested for phenylketonuria, which is an inborn metabolic error of metabolism that results in mental retardation, whereas testing for muscular dystrophy (MD) is not routine. Yet, MD in males is reported in some studies as occurring two or three times more frequently than does PKU. A

screening test is available for MD, and we can confirm the diagnosis in newborn babies. Why, then, is the screening test for MD not routine, as is the test for PKU? Part of the answer may be that MD is, at present, untreatable. Many people, including medical professionals, are not ready to accept screening for a disorder that is not treatable. Other arguments against screening are the cost of the procedure and the inaccuracy of the test. Some children who are normal will be diagnosed initially as having MD, which requires that they undergo additional testing and causes considerable anxiety in the parents. MD may not show up in some boys until the age of three or so, but it is still fatal by about age twenty. Some advocates justify MD screening on the basis of the three-year wait. A couple could have one or even two more affected males before they found out that a son had the disease. The prenatal test for MD would provide those families with an additional option in future pregnancies.

A complex series of problems has emerged from the widespread screening of all pregnant women for the prenatal detection of spina bifida and chromosomal abnormalities, mainly Down syndrome. Those issues are discussed on page 71 in the article "Prenatal Diagnosis," in the section on maternal serum screening.

SCREENING ON THE JOB

Recently, genetic screening has moved into the workplace. For example, there is evidence that large numbers of people have an inherited enzyme deficiency that makes their lungs susceptible to damage from cigarette smoke, asbestos particles, and other atmospheric pollutants. Those indi-

viduals tend to develop chronic emphysema, which is a loss of elasticity in the lungs that considerably shortens their lives. The managers of an asbestos mining company, after learning that it was possible to screen employees and prospective employees for this genetic predisposition, instituted company-wide screening. They planned to identify susceptible miners and retrain them for other jobs in the company. New employees were to be screened so that susceptible applicants would not be hired. The union responded by taking legal action against the company, charging discriminatory labor practices.

Genetic screening, like many advances in science and technology, often creates new problems as it solves others. Some of the more obvious and difficult questions include the following:

1. To what extent should confidentiality be protected? For example, should the relatives of a carrier be told about the carrier status against the wishes of the individual?

2. What uses of information obtained from screening might constitute violations of human rights? What practices or procedures might we view as discrimination against individuals, families, or groups?

3. How much should society be permitted to do for a person's "own good"? On what grounds might laws be passed to limit individual choice in order to obtain some "public good"?

4. Is it just or unjust to withhold screening services from particular individuals or groups? Why? Is it just or unjust to withhold the information obtained? Why?

THE SCREENING OF BENJAMIN MILLER

My name is Ruth Shapiro. I teach tenth-grade biology at Park Hill High School in Lincoln, Nebraska. My story is about what I learned from Greg Miller, one of my students.

His new baby brother, Benjamin, was born about six weeks ago, and for two weeks Greg had been beaming. Then, about a month ago, he seemed terribly anxious about something. For two or three weeks, he seemed nervous, distracted, and unable to concentrate. But one Tuesday, I was relieved to see Greg happy and smiling in class.

At lunch I joined the group of students around Greg. I asked Greg how Benjamin was doing. "Well, he has, he has, has. . . ." Greg became obviously upset again and seemed on the verge of tears. I put my arm around him and said gently, "You don't have to talk about it if you don't want to, Greg. But I've noticed that lately you've been upset about something. I've been wondering if it is anything I can help you with."

With that, the dam burst, and Greg told me the whole story. When Benjamin was born everything was fine, and he and his mom were home from the hospital in three days. But about a week later, Greg's mom got a call from the pediatrician, who said that Benjamin had an abnormality on his newborn screening test and needed to come into the hospital. Greg said the next day Benjamin and his mom went to Children's Hospital in Omaha and were there for three days. When they got back, his mom seemed nervous but said everything with Benjamin was fine. He just needed to be on a special baby formula. But, Greg kept hearing his parents talk about PKU and mental retardation. Greg became scared. He was afraid to ask his parents too many questions, because he thought they had enough to worry about.

But last weekend, Greg, Benjamin, and their parents had spent a day with the Lemansky family in Kearney. It turned out they have a boy one year older than Greg who also has PKU, and the whole Miller family had learned a lot and felt much better about the disorder.

That night, I called Greg's father to tell him about my conversation with Greg and to offer to help in any way I could. He said Mr. Lemansky was a life science instructor at Carroll Community College. He suggested I call Mr. Lemansky. I did, and he volunteered immediately to talk to Greg's class about genetic screening. He came one week later. I recorded his talk, and here is a shortened version:

"When our first son, John, was born we were very happy. He appeared perfectly normal. When he was a little older, we marveled at his exceedingly light coloring. His blond hair was almost white. And his eyes were the bluest blue you've ever seen. His mother also noted that he had an unusual odor about him. I really didn't notice it, but it upset her a great deal. John was a healthy boy, but his development was slow. He didn't sit up alone until he was a year old, and he didn't walk until he was two.

He learned to say 'mama' and 'dada,' but not much more than that.

"Our doctor kept reassuring us that John was normal— 'only a bit slow.' And to make sure, he did a thyroid-function test, with normal results. I was in graduate school in those days, and so poor we could hardly pay the doctor's bills. After five years, it finally occurred to me to take advantage of the university's family health service. A young pediatrician took one look at John, did a urine test, and made the diagnosis of PKU.

"The letters PKU stand for *phenylketonuria*, an autosomal-recessive, inborn error of metabolism discovered by Folling in Sweden in 1934. The metabolic defect is a deficiency of the enzyme phenylalanine hydroxylase. Normally, that enzyme converts the amino acid phenylalanine to another amino acid, tyrosine. When the conversion doesn't happen, excess phenylalanine accumulates in the blood to levels ten times greater than normal. The result is phenylalanine 'poisoning.'

"In the overwhelming majority of cases, that leads to severe mental retardation and all the other symptoms that occurred in our son. These include the fair coloring and unusual odor and also skin rash and convulsions.

"As I mentioned, PKU is a recessively inherited condition. Our risk of having another child with PKU is twenty-five percent with each pregnancy. Prenatal diagnosis wasn't available fifteen years ago, but there was a screening test to detect the disorder at birth. Treatment was possible, also. We desperately wanted another child, and we had Kevin. The doctors told us shortly after he was born that Kevin also has PKU, but they got him started on the treatment immediately. Treatment is possible because scientists have developed a special diet. Its mainstay is a formula made out of milk protein components from which the manufacturer removes the phenylalanine with charcoal.

"The diet is easy to maintain during the bottle-feeding period, but difficulties arise later on. Keeping Kevin on the special diet requires our constant guidance, but he appreciates the importance of the diet. The diet is not attractive or very tasty. A low phenylalanine diet usually means no meat, fish, cheese, eggs, normal bread, or cake. Instead, Kevin eats small amounts of green vegetables, potatoes, and fruits. A specially prepared milk substitute, low protein foods, and special cornstarch products constitute most of the diet. One particular problem is that, as yet, there is no good-tasting bread substitute. Children with PKU must follow the complex diet plan. The child remains on the diet during the years the brain is still growing, or until at least age ten. Scientists are now studying whether children can stop the diet at ten years old or whether they should continue, even into adult life. Available evidence indicates that people treated in such a manner have normal intelligence. The diet prevents the symptoms and complications of PKU.

"Problems arise when a successfully treated woman with PKU wants to become pregnant. In the more than one hundred offspring of such women studied, almost all are retarded and have minor anomalies. About one-fourth have major malformations, as well. A homozygous fetus of a carrier mother is protected from prenatal brain damage because its mother clears the fetal bloodstream of excess phenylalanine. But, in the case of the homozygous *mother*, all of whose fetuses will at least be heterozygous, the placenta maintains a higher level of phenylalanine in the fetus than in the mother. A fetus exposed to these high levels of phenylalanine during the entire nine months of pregnancy will be born with brain damage. To prevent that, the mother must be on a low-phenylalanine diet during the pregnancy. Distasteful as the diet may be, homozygous women contemplating pregnancy have to go back on the diet some time before conception occurs.

"For the diet to be effective in a newborn with PKU, it should be started within the first two weeks of life. How do doctors determine who needs the diet? They do that by screening all newborn babies for this genetic disorder. What do we mean by genetic screening?

"Genetic screening is a search in a population for people with certain genotypes. These genotypes are (1) those already associated with disease or predisposition to disease, (2) those that may lead to disease in descendants, or (3) those that produce other variations not known to be associated with disease. PKU clearly falls into the first category, because successful detection through screening leads to effective treatment and genetic counseling to prevent recurrence. But what conditions must exist before a new screening program is implemented on a statewide or regional basis?

"A committee of the National Academy of Sciences in Washington, D.C., reported that genetic screening, when carried out under controlled conditions, is an appropriate form of medical care when it meets the following criteria:

- There is evidence of substantial public benefit and acceptance, including acceptance by medical practitioners.
- The feasibility of screening has been investigated and it has been found that benefits outweigh costs; appropriate public education can be carried out; test methods are satisfactory; laboratory facilities are available; and resources exist to deal with counseling, follow-up, and other consequences of testing.
- An investigative pretest of the program has shown that costs are acceptable; education is effective; informed consent is feasible; aims of the program with regard to size of the sample to be screened, the ages of the screenees, and the setting in which the testing is to be done have been defined; laboratory facilities have been shown to fulfill requirements for quality control; techniques for communicating results are workable; qualified and effective counselors are available in sufficient numbers; and adequate provision for effective services has been made.
- The means are available to evaluate the effectiveness and success of each step in the process.

"The committee published its report in 1975. Screening for PKU began some ten years before that, however. Screening began before researchers could thoroughly test the validity and effectiveness of all aspects of treatment, including appropriate dietary treatment. Nevertheless, current assessment of those screening programs shows that the methods are reasonably efficient, the means for moving from test to treatment are adequate, and the appropriate dietary treatment is harmless and effective. Now, all states screen for PKU.

"What happens in actual practice? Since blood phenylalanine levels rise only very slowly in a newborn, it's best to test the infant several days after birth. The doctor takes a drop of blood and puts it on a special kind of filter paper. These filter papers are sent to a testing lab where automated equipment handles many specimens at once.

"The test is quite ingenious. Dr. Robert Guthrie of New York developed it. The test is based on an increase in bacterial growth caused by excess phenylalanine. Promotion of bacterial growth on the filter paper indicates an excess of phenylalanine in the infant's blood. When that happens, the lab notifies the physician, and takes a more precise measurement of the blood phenylalanine level. If the blood level is high, we presume the infant has PKU. The doctor begins giving the child the special diet. After the blood phenylalanine level drops to a normal range, the diet is stopped. The doctor then gives the infant a large amount of phenylalanine to see how he or she responds. If the level of phenylalanine rises and the level of tyrosine drops, the diagnosis is considered confirmed. The child continues on the treatment program.

"For PKU, the cost of the test is small, and the treatment is highly effective. On the other hand, the cost of caring for a severely affected individual in an institution for many years is catastrophically high, both in emotional and financial terms.

"Most states screen for PKU along with a test for hypothyroidism, a nongenetic cause of metal retardation. The average cost for the tests is about twelve dollars. For every one hundred thousand infants tested, about thirty-five will have either PKU or hypothyroidism. Those infants will develop retardation if they are not detected and treated. Treatment for these conditions can last about twenty years. The cost for treating all thirty-five people for twenty years would be about $540,000.

"If we did not detect those thirty-five infants with PKU or hypothyroidism, they would become retarded. The institutional care for those thirty-five retarded people would cost about $4,900,000 during the course of their lives. So, the costs and benefits for screening work out like this."

Mr. Lemansky wrote these figures on the chalkboard:

Costs for screening and treatment (per 100,000 infants)

each screening test = $12

$12 x 100,000 infants = $1,200,000

treatment for 35 people for 20 years = $540,000

total cost to screen and treat = $1,740,000

Costs for care of 35 retarded people if not detected = $4,900,000

$4,900,000
–$1,700,000
$3,200,000 saved per 100,000 infants screened

$3,200,000 ÷ 35 people = $91,428 saved per person detected

"So," Mr. Lemansky continued, "you can see that screening for PKU and hypothyroidism is cost effective. It leads to prevention through correct diagnosis, early treatment, and genetic counseling. Kevin has been on treatment for fourteen years, and as Greg can tell you, he is as normal as any fourteen-year-old.

"I maintain my diet," Kevin said, "and I'm used to not eating all the foods my friends do. I admit I cheat on my diet sometimes, but not often, because I know how important it is and how good it feels to be on it. My older brother, John, is now in an institution for the severely retarded. I'm glad that didn't happen to me."

"Besides widespread newborn screening," Mr. Lemansky continued, "there also is now a prenatal test available for PKU. The disorder is detectable using amniocentesis or chorionic villus sampling. Prenatal testing is available to families who have already had one child with PKU and so have a twenty-five percent risk for having another. I understand from your teacher that you have an article on prenatal diagnosis in your genetics magazine where you can go to learn more about the various procedures."

During our class break, we talked more about genetic screening and shared some snacks that Greg had brought in honor of Benjamin's birth.

REFERENCES

Andrews, L.B., J.E. Fullarton, N.A. Holtzman, A.G. Motulsky, eds. 1994. *Assessing Genetic Risks: Implications for Health and Social Policy.* Washington, D.C.: National Academy Press.

Clayton, E.W. et al. 1996. "Lack of Interest by Nonpregnant Couples in Population-based Cystic Fibrosis Carrier Screening." *American Journal of Human Genetics* 58:617–627.

Committee for the Study of Inborn Errors of Metabolism (National Research Council). 1975. *Genetic Screening: Programs, Principles, and Research.* Washington, D.C.: National Academy of Sciences.

Holtzman, N.A. 1989. *Proceed with Caution: Predicting Genetic Risks in the Recombinant DNA Era.* Baltimore: Johns Hopkins Press.

McCabe, L. 1996. "Efficacy of a Targeted Genetic Screening Program for Adolescents." *American Journal of Human Genetics* 59:762–763.

Motulsky, A.G. 1997. "Screening for Genetic Disorders." *New England Journal of Medicine* 336(18):1314–1316.

Nelkin, D., and L. Tancredi. 1989. *Dangerous Diagnostics.* New York: Basic Books.

Office of Technology Assessment. 1988. *Healthy Children.* Washington, D.C.: U.S. Government Printing Office.

THE BOY WITH THE EXTRA X CHROMOSOME

Dr. Monica Row, a biology teacher at Jefferson High School, was using the copy machine in the school office. She saw Tom Benes, one of her tenth-grade students, waiting with two dimes and a letter to be copied. Dr. Row and Tom had become good friends. Tom had done a special project in genetics earlier in the year, and in addition, Dr. Row was the school bandmaster, and Tom played clarinet. Dr. Row had sensed for several weeks that something was bothering Tom.

"Let's have it, Tom; I'll copy it for you," offered Dr. Row. Tom declined the offer, but did not leave.

"Can I talk to you about something?" asked Tom.

"Sure, Tom. Can you come to my classroom during lunch hour?" Dr. Row suggested. Tom agreed.

Tom showed up at noon, but just picked at his lunch. He looked ill at ease. He was a bright student and the best clarinet player the school had ever had. Tom had been seriously considering dropping out of the band, however, because he was self-conscious about his height. At six feet four inches (183 cm), he towered over everyone, and he felt that his band

uniform fit poorly. To make matters worse, in spite of his height, Tom appeared immature. He had no beginnings of a beard and his voice was still high. He avoided sports because he was embarrassed about his undeveloped genitals.

The letter Tom had was from a pediatrician who specializes in inherited diseases. The Benes' family doctor had referred Tom and his parents for genetic counseling. The letter on page 89 is from the Genetics Clinic, Community General Hospital, East Rivertown, Ohio.

By the time his teacher had finished reading the letter, Tom felt a little more comfortable. "Dr. Row," he said, "I brought the letter to you mainly to thank you for making our sessions last fall on the sex chromosomes so clear and interesting. When the geneticist started talking about Klinefelter syndrome, I started to feel really bad. I felt embarrassed and upset until Dr. Ramirez started to talk about X inactivation and the Lyon hypothesis. I couldn't believe that I could remember anything when I was so nervous, but hearing something I knew about sure helped relieve the tension!"

"Tom, I'm delighted to hear that our work together helped you. Even from a quick look at this letter, I have learned a great deal. Tell me more about genetic counseling. I'm particularly interested in the presence of the medical students. Did they just appear with the doctor, or did you and your parents have some say in the matter?"

"I'm glad you asked that," Tom replied. "Before we went into the geneticist's office for the counseling, we talked quite a bit with her about what a new subject medical genetics is and how important it is for doctors to understand the new information about genetics and to understand how people react to learning about a genetic disorder in the family. She said right off that she understood how nervous we were and that it would be perfectly okay if we decided not to have any medical students attend our session. Then, she gave my parents and me all the time we needed to decide. Although I nearly freaked later when she started talking about such personal subjects—you know, one of those students was a woman and she didn't look much older than I am—the geneticist made things so interesting with pictures of chromosomes and by having everyone participate in the discussion, that before long we were *almost* relaxed."

"Tom, you said you wanted to thank me and that's very gratifying, but is there anything else on your mind?"

"Yes, there is. I know I can call Dr. Ramirez or even go to see her again, but I have a couple of questions that I bet you can answer. I read in a textbook that the extra X chromosome causes mental retardation and obesity. Are those things going to happen to me?"

"Well, that's a question you should ask Dr. Ramirez, but from my reading on the subject, I'd say definitely not. Among those few 47, XXY men who are mentally retarded, the retardation is present at birth and does not get worse as the person grows older. With regard to becoming overweight, every human being, no matter how many X chromosomes he or she possesses, has a genetically determined predisposition to accumulate varying amounts of fat. That predisposition even varies at different times in our lives. With only rare exceptions, we can control our genetic predisposition through diet and exercise. I think you will be careful about your food intake and get enough exercise to keep your weight in the normal range."

Tom then asked some fundamental questions about gene action in the X chromosome. "If one of the two X chromosomes is inactivated in normal 46, XX females, why don't all females have at least some of the features of the 45, X Turner syndrome (short stature, webbed neck, absence of secondary sexual characteristics, sterility, and other problems)? Along the same line, why do males with two or more X chromosomes have any problems at all?" Tom also wondered why the textbooks seem to indicate that most 47, XXY males have many physical problems and are mentally retarded, when we now know that, except for infertility, most are quite normal. Finally, he got up enough nerve to ask Dr. Row about artificial insemination. Is it available? Is it legal? Where do the donors come from? How do you know they're normal?

Being a recipient of genetic counseling and knowing that he had an extra X chromosome could not change Tom's condition. But within a few months his beard and pubic hair began to appear. Understanding his problem helped him regain and further develop his self-confidence. When he stopped slouching, his band uniform fit him perfectly.

Dear Mr. and Mrs. Benes and Tom:

This is the letter we promised to send following your visit to our clinic for genetic counseling last Monday. As you will recall, I told you that we prepare letters like this for most of the patients and families referred to us. Genetic conditions tend to be unfamiliar to most people. A letter like this usually helps our clients remember what we talked about. In addition, the copy we sent to your family physician, Hugh Johnson, will serve as a record of the consultation and as a source of information should you wish to discuss any of these things with him in the future. Of course, we also have sent Dr. Johnson the complete test results.

I will begin by reviewing the events that led you to our genetics clinic in the first place. Tom, you have been and still are a healthy sixteen-year-old student doing well academically and socially. All three of you became a bit concerned when Tom went through the usual teenage growth spurt without showing any signs of voice deepening, beard development, or hair in the armpits and groin. At first, Dr. Johnson—quite correctly—was not concerned, since there's such a wide variation in the age at which sexual maturation occurs in both boys and girls. When he examined Tom, however, he noted that his testicles were smaller than expected for this stage of his development. For that reason, he ordered some tests to evaluate the level of male sex hormone in Tom's blood. He also ordered a chromosome analysis. When I phoned to give your doctor the results of the chromosome test, he decided to recommend that the three of you come to see us for genetic counseling.

Tom, I'm sure you recall my telling you that you are one of many individuals who has one more X chromosome than usual. Chromosomes are the structures in which genes are located. Each person has thousands of genes, usually in forty-six chromosomes. Both our genes and our chromosomes come in pairs, one set from our mothers and one from our fathers. One pair of chromosomes, the X and Y pair, determines our sex. Because of an error in meiosis that occurred either in the egg or the sperm from which you developed, Tom, you ended up with that extra X.

About one out of every thousand males has this 47, XXY chromosome constitution, instead of the usual 46, XY. Geneticists refer to the condition as Klinefelter syndrome and list a number of characteristics that may be associated with it. These include tall stature, some minor birth defects, small testicles and sterility, and even mental retardation. Obviously, the most serious of those do not apply to you. Nor, in fact, do they apply to most of the many males in the world who are also 47, XXY. As best we can tell, the extra X chromosome does relatively little harm because it is inactivated. The idea that one X chromosome is inactivated is called the Lyon hypothesis.

Most men who are 47, XXY appear perfectly normal mentally and physically and many never find out they have an extra X. The only consistent findings are sterility—due to no sperm production—and a tendency to be tall. Nearly all 47, XXY men mature sexually without any hormone treatments and most become sexually active.

At our meeting, we discussed marriage, infertility, and possible options for having a family, including adoption and artificial insemination. I believe it was you, Mrs. Benes, who wondered whether this obviously "genetic" condition was "hereditary." That detailed family history we took, on both you and Mr. Benes, failed to turn up any relatives with problems we might associate with 47, XXY or any other chromosomal disorder. That was as we expected, since we have no examples of more than one individual who is 47, XXY in a family. I did mention, however, that there are families with an apparent tendency to repeated nondisjunction during meiosis—for example, a child with 47, XXY and another with Down syndrome. Were you and Mr. Benes to have another child, you might wish to consider amniocentesis.

I hope our counseling session and this letter have answered your questions. If not, please don't hesitate to contact me. I'd be happy to see you any time. Also, thank you for allowing the two medical students and the pediatric resident to join us. It was a valuable learning experience for them. They asked me to include their thanks in my letter.

Yours sincerely,

Robin Ramirez, MD

CAREERS IN GENETICS

Genetics is a rapidly growing branch of biology. It offers many and varied career opportunities now and will continue to do so in the future.

AGRICULTURAL GENETICS

The most impressive early achievements of genetics were in plant and animal breeding. Those accomplishments led to tremendous improvements in crop productivity and in livestock. Because serious food shortages occur in many areas of the world, improvement in the productivity of crops is a concern of biologists. As the world's population continues to grow, food shortages are likely to become more severe.

Agricultural geneticists work to develop new plant strains that are more resistant to pests and disease, higher in nutritional value, and capable of higher yields. Newly developed strains yield billions of dollars in cash crops each year. Selective breeding of cows and other livestock has produced animals that are more resistant to insect diseases and capable of producing a better quality and quantity of meat.

Although genetic research has improved the quality and yield of crops and livestock, we need to work even harder to meet the world's ever-increasing demand for food. Many state universities have schools and colleges of agriculture. If you are interested in agricultural genetics as a career, contact one or more such schools for more information.

HUMAN GENETICS

Human genetics is the study of human variation. Sometimes, we include normal human variation under the field of physical anthropology. In the U.S., anthropology is now separate from genetics. In certain European countries, however, professors of human genetics are also professors of anthropology, maintaining a close link between the two.

The study of normal human variation has only begun. Hundreds of different human population groups await closer characterization of physical attributes. We need more information about such things as height, head size and shape, functional characteristics (strength, endurance, lung capac-

ity, and liver functions), blood groups, serum proteins, enzymes, and other characteristics.

You can improve your career opportunities in both agricultural and human genetics through graduate studies leading to master's or doctoral degrees. Additional postgraduate study in a special field also is an advantage.

MEDICAL GENETICS

Medical genetics is the study of abnormal human variation. This includes conditions due to chromosomal anomalies, multifactorial gene-environment interaction, and mutations. New methods of chromosome analysis have led to the discovery of hundreds of additional syndromes that result from structural changes in chromosomes. Most genetic conditions are multifactorial disorders, just as most normal characteristics are multifactorial. Those disorders account for most birth defects, as well as a large number of diseases of adults.

We expect other areas of human genetics to expand consistently in this century and the next. These include immunogenetics, biochemical genetics, pharmacogenetics, and behavioral genetics.

For those wanting to pursue a career of research in these fields, we recommend graduate study toward a PhD and postdoctoral studies. For those interested in clinical genetics—that is, using the practice of medicine to help people with genetic defects—we recommend medical school and post-MD training, including residency and postdoctoral fellowship.

GENETIC COUNSELING

During the last twenty-five years, a new group of professionals has arisen in response to the enormous demand for genetic services. These are the genetic counselors. Many genetic counselors have master's degrees with special training in genetic counseling. Some are physicians and some have PhD degrees in genetics. They provide a wide range of services in clinics and hospitals. Demand for well-trained and experienced genetic counselors is outstripping supply. And, the demand probably will increase. Genetic counselors now serve on the staffs of institutions for the retarded, blind, and deaf. They work in schools and in state offices such as bureaus of maternal and child health. They are employed in city and county nursing and social service departments, large hospitals, and many other institutions.

PROFESSIONAL QUALIFICATIONS

The American Board of Medical Genetics and the American Board of Genetic Counseling assure high standards of training and knowledge for those who work in clinical genetics or genetic counseling. Board examinations are given to four groups: genetic counselors with master's degrees; clinical geneticists who are physicians; PhD human geneticists; and PhD (and/or MD) laboratory geneticists. This last group includes cytogeneticists, biochemical geneticists, and immunogeneticists. All of those groups must keep up with changes and new knowledge in their fields. The U.S. plan is similar to the plan already established in Canada.

OTHER CAREER OPPORTUNITIES

If you prefer not to go beyond a four-year college education, there are many other career opportunities—particularly as trained and certified laboratory technicians to work in immunogenetic, biochemical, and cytogenetic labs. Those who are interested in careers in genetics should contact local hospitals, clinics, colleges, universities, or the following organizations for information.

American Society of Human
  Genetics
9650 Rockville Pike
Bethesda, MD 20814
Phone: (301) 571-1825
FAX: (301) 530-7079
http://www.faseb.org/genetics

Genetics Society of America
9650 Rockville Pike
Bethesda, MD 20814

Phone: (301) 571-1825
FAX: (301) 530-7079
http://www.faseb.org/genetics/
  gsa/gsa-int.htm

National Society of Genetic
  Counselors, Inc.
233 Canterbury Drive
Wallingford, PA 19086-6617
Phone: (610) 872-7608
FAX: (610) 872-1192
E-mail: nsgc@aol.com
http://members.aol.com/
  nsgcweb/nsgchome.htm

A NEW VIEW OF THE DISABLED

PERSONAL ATTITUDES

How we view a person with a disability or disfigurement is largely determined by a phenomenon we call spread. Spread is how a disabled person's disability is perceived to affect—or "spread"—to other characteristics of that person.

Beatrice Wright, a professor of psychology at the University of Kansas, and other psychologists have gathered evidence on spread. They have found, for example, that our attitudes toward blindness as a condition tend to be more negative than are our attitudes toward blind people as individuals. Dr. Wright suggests that the way we view and react to people with disabilities is strongly influenced by two vastly different attitudes or frameworks. Dr. Wright calls these coping and succumbing attitudes (see Table 1).

Briefly, the succumbing attitude tends to highlight the difficulties and tragedy of being disabled, to

TABLE 1 Contrasts between coping and succumbing attitudes.	
Coping	Succumbing
1. The emphasis is on what the person *can* do.	1. The emphasis is on what the person *can't* do.
2. Areas of life in which the person can participate are seen as worthwhile.	2. Little weight is given to the areas of life in which the person can participate.
3. The person is perceived as playing an <u>active</u> role in molding his or her life constructively.	3. The person is seen as <u>*passive,*</u> as beaten down by difficulties.
4. The accomplishments of the person are appreciated in terms of their benefits to the person and others (asset or intrinsic evaluation), and not primarily devalued because they fall short of some irrelevant standard.	4. The person's accomplishments are minimized by highlighting their shortcomings (comparative evaluation, usually measured in terms of "normal" standards).
5. The negative aspect of the person's life, such as the pain that is suffered or difficulties that exist, are felt to be manageable. They are also seen as limited because satisfactory aspects of the person's life are emphasized.	5. The negative aspects of the person's life, such as the pain that is suffered or difficulties that exist, are kept in the focus of attention. They are emphasized and exaggerated and seen to usurp all of life (spread).

emphasize what the person can't do. The coping attitude, on the other hand, enables one to seek solutions to problems. We see people with disabilities as actively attempting to lead their lives constructively, and not as passively destroyed by their limitations and difficulties. The coping attitude should not make us callous to the hardships of disabled persons. Instead, it suggests that suffering has its limits and that despair is not a necessary outcome of the disabling condition.

PUBLIC RESPONSIBILITY

To begin to overcome the prevailing succumbing framework, we must alter the way we speak of disabled persons, for a person is not equivalent to the impairment. Labels like *"the* deaf," *"the* mentally retarded," and *"the* wheelchair patient,"* tend to cause us to consider all persons with similar disabilities as a group, as if such persons are all the same. As we have seen, that is false at the genetic level and is no less false in the public and social context.

Public responsibility toward the disabled is a topic of much debate. Recent events reveal that change may be in the wind. Modifications of buildings are now commonplace. Discriminatory practices, such as refusing to employ a disabled person when the person can do the job, are illegal. And suits have been filed against those who have allegedly denied housing to the disabled. Such shifts in public policy may indicate a more widespread acceptance of the coping attitude than used to be the case. But the lines

are not clearly drawn, as the following case study illustrates.

Charlene was born with various physical abnormalities. She is a low-birth-weight dwarf. Her eyes are, it seems, too large, and her nose is beaklike. Her facial features are unusual; her face is narrow and her lower jaw recedes. In addition, her right foot is twisted.

The medical geneticist called in consultation recognized those conditions as a specific form of dwarfism inherited as an autosomal recessive trait. In taking the family history, the geneticist learned that Charlene's parents are first cousins. Although Charlene's life expectancy is not likely to be shortened, the geneticist pointed out that all known patients with this condition are profoundly mentally retarded. Institutionalization probably is necessary.

Charlene's parents are of modest means. Charlene's father, recalling the severely retarded brother of his best friend in high school and what he termed the "devastating impact" on that family, had already decided that Charlene would not be cared for at home.

Charlene's parents live in a state that does not provide any appreciable funds for the care of the retarded in institutions. Various legal actions are in progress that might require institutions to make improvements, but, generally speaking, these institutions are destined to remain inadequate in their provision of reasonable care to retarded residents.

From this case, we can raise some important questions:

1. Do health-care providers, doctors and nurses, tend to view people like Charlene with succumbing or coping attitudes?
2. What attitude did Charlene's father display?
3. Should Charlene's parents feel guilty, since conditions inherited as autosomal recessive traits occur more frequently as a result of consanguineous (blood-related) marriages?
4. What is our public responsibility for providing institutional care for retarded persons like Charlene?
5. How should a state decide what resources it should provide for the care of people like Charlene? Is money a serious problem?
6. In our society, questions about public responsibility often translate into questions of taxation and allocation of tax revenues. How would you, as a citizen at a public meeting at Town Hall, argue for additional tax money to spend on institutional care for people like Charlene? How do you think people would react if you presented Charlene as a victim of her disability? How would they respond if you portrayed Charlene as competent and productive?
7. Suppose that, instead of institutional care, you wanted tax money to spend on education for people like Charlene. Would your approach differ? Why or why not? What are the long-range consequences of that approach?

EXCEPTIONS TO MENDEL'S LAWS

It often has been said that Mendel was a lucky biologist. He chose as his experimental model a relatively simple plant, the garden pea, and he also chose traits that were either unlinked (on different chromosomes) or sufficiently far apart on the same chromosome that it didn't matter. His traits segregated neatly, although there are other traits in the pea that don't. We will never know whether he chose wisely or was just lucky. But, we do know that he was a meticulous investigator and that he was quite familiar with the earlier experiments of other naturalists along similar lines, so we probably would guess the former. Keep in mind that Mendel knew nothing about chromosomes or genes. He postulated the existence of what he called hereditary factors, obviously what we now know as genes. For more on Mendel and his work, see "Genetics in History: Gregor Mendel."

In this section, we will look into traits that do not follow Mendel's laws, not in any way to diminish the importance of his observations and conclusions. In fact, without his basic laws of heredity, the exceptions would not have been recognizable and progress in the understanding of genetics would have been greatly delayed.

X-LINKED INHERITANCE

X-linked recessive inheritance was the first observed exception to Mendel's laws—we see the trait or disorder only, or primarily, in males. This article will take you from the first observation of this unexpected phenomenon in male fruit flies shortly after the turn of the twentieth century to the modern molecular theory that explains how one X-linked gene changes as it passes from generation to generation in human families and causes progressively more severe symptoms.

THE SEX CHROMOSOMES

Human cells have twenty-three pairs of chromosomes—twenty-two pairs of autosomes, and one pair of sex chromosomes. In human males, the sex chromosomes are identified as the X and Y chromosomes. The Y chromosome is much smaller than the X chromosome. Human males produce two kinds of sperm. Half carry an X chromosome and half carry a Y chromosome. In human females, sex chromosomes are both Xs. All normal eggs produced by a female carry one X chromosome. If an egg is fertilized by an X-bearing sperm, the offspring is a female (XX). If an egg is fertilized by a Y-bearing sperm, the offspring is a male (XY).

Not all animals have sex-determining sperm. In some species, such as birds, butterflies, and moths, the female has two different types of sex chromosomes and produces two kinds of eggs. In those organisms, all sperm carry the same kind of sex chromosome. The chromosome present in the egg, therefore, determines the sex of the offspring.

In still other species, one of the sex chromosomes may be absent. In some grasshoppers, for instance, males have twenty-two autosomes (eleven pairs) and one X chromosome. Females also have twenty-two autosomes, but they have two X chromosomes. Half the sperm have eleven autosomes. The other half have eleven autosomes and one X. The sex of the young grasshopper depends on the kind of sperm that fertilizes the egg. In this case, the presence of one X chromosome results in a male grasshopper.

In most plant species, the same plant produces both male and female sex cells. We find sex chromosomes, however, in plant species that do have separate male and female individuals. Such plants include holly, asparagus, willows, and cottonwoods.

THE DISCOVERY OF X-LINKED INHERITANCE

Around 1910, Thomas Hunt Morgan was raising thousands of fruit flies in the laboratories of Columbia University, in New York City. As far as anyone knew at that time, fruit flies found in nature had dark red eyes. While examining fruit-fly cultures, however, Morgan found a male fly with white eyes. When Morgan mated that male fly with a red-eyed female, all of the F_1 generation had red eyes. That was not surprising, because Morgan assumed that flies inherited white eyes as a recessive trait.

Morgan then mated flies of the F_1 generation to each other. The F_2 generation showed a ratio of three-quarters red-eyed flies to one-quarter white-eyed flies. That ratio of eye colors also was expected. But, all the white-eyed

flies were males! That trait seemed to be linked to the sex chromosomes. Further experiments confirmed that finding.

White eyes in fruit flies are an X-linked trait. Morgan's knowledge of meiosis led him to conclude that male fruit flies produce two kinds of sperm. Half of the sperm would carry an X chromosome and half would carry a Y chromosome. Each sperm also would carry one chromosome from each pair of autosomes. Females, on the other hand, would carry one X chromosome, plus one chromosome from each pair of autosomes. Apparently, the Y chromosome was inactive in the inheritance of eye color. Thus, in the fruit fly, the gene for eye color is on the X chromosome.

HUMANS ALSO HAVE X-LINKED TRAITS

Humans have a pair of sex chromosomes similar to the sex chromosomes of fruit flies, so we

might expect that some human hereditary traits would be X-linked. We now have identified a number of X-linked traits in humans, including some types of colorblindness, hemophilia, and muscular dystrophy. Of those X-linked traits, colorblindness is the most frequent, occurring in eight percent of white males.

True colorblindness—that is, the inability to see any of the primary colors—is common in nocturnal animals, which are active only at night. In humans, true colorblindness is extremely rare. Many people who are called colorblind might better be called "color weak." These individuals confuse reds, greens, yellows, and blues. The most common form of colorblindness is red-green, which occurs in about five percent of white males. The distinctive patterns of inheritance associated with X-linked traits are apparent in pedigrees such as those shown in Figure 1 for red-green colorblindness.

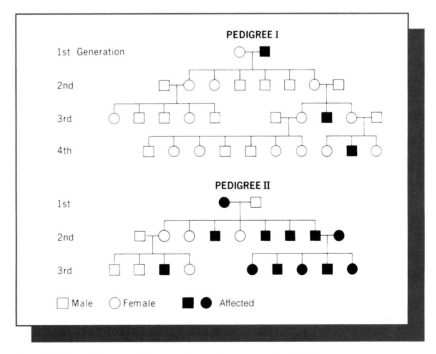

FIGURE 1 ▪ Two pedigrees for red-green colorblindness.

Study of those pedigrees should allow you to answer such questions as:

1. Is colorblindness a dominant or recessive trait? Use pedigree I to support your answer.
2. Is colorblindness X-linked? Use pedigree II to support your answer.

   After you have answered those two questions, predict the expected results of a mating between a man who is color-blind and a woman who is homozygous for normal color vision.

3. Among their male offspring, what fraction would have normal vision?
4. What fraction would be color-blind?
5. Among their female offspring, what fraction would have normal color vision?
6. What fraction would be carriers?
7. What fraction would be color-blind?

   Repeat the calculations, this time for a mating between a carrier woman and a man with normal color vision.

8. Among their male offspring, what fraction would have normal color vision?
9. What fraction would be color-blind?
10. Among their female offspring, what fraction would have normal color vision?
11. What fraction would be carriers?
12. What fraction would be color-blind?

More men are color-blind than women. It is not difficult to see why. If a man receives an X chromosome that carries the gene for colorblindness from his mother, he will be color-blind. That is because there is no allele on the Y chromosome to offset the allele for colorblindness on the X chromosome. For a woman to be color-blind, however, she must receive two X chromosomes that carry the allele for colorblindness, one from her mother and one from her father. Therefore, her mother would have to be a carrier and her father would have to be color-blind—an infrequent combination.

If one X chromosome in ten were to carry the allele for colorblindness, one male in ten would be color-blind. To find out how often females would be color-blind, apply the second principle of probability. This principle states: the chance of two independent events occurring together is the product of their chances of occurring separately. The chance of a female having two X chromosomes, each carrying the gene for colorblindness, would be the product of the two separate probabilities, .1 x .1, or .01. Thus, the frequency of color-blind women would be only about one-tenth of the frequency of color-blind men (1 in 100 for the female compared to 1 in 10 for the male). The actual frequency of color-blind men ranges from five percent to nine percent in different populations.

The most common type of hemophilia is also an X-linked trait in humans. It is a much more serious condition than colorblindness. Fortunately, it is rare. In hemophilia, either the blood fails to clot after an injury, or it clots very slowly. Some males who have hemophilia have frequent, and painful, bleeding around the joints. Others are more mildly affected; they bruise easily and bleed extensively if their skin is cut.

X-linked hemophilia has a royal history. The trait appeared in a son and three grandsons of Queen Victoria of Great Britain. Because of frequent intermarriages among the royal families of Europe, the gene spread widely. During the nineteenth and early twentieth centuries, the gene for hemophilia affected the course of history, especially in Spain and Russia. For example, Rasputin, a Russian Orthodox monk and mystic, claimed to be able to control the bleeding in young Prince Alexis Czarevich, a hemophiliac. (The name Czarevich indicates that Alexis was the eldest son and heir of Czar Nicholas II, who was the last Czar of Russia.) The Czar's fear of Rasputin's power apparently affected his decisions in certain matters of state. Figure 2 shows a pedigree of the distribution of hemophilia in Queen Victoria's descendants.

We have used colorblindness and hemophilia as typical examples of X-linked inheritance for many years. But, more than two hundred other X-linked conditions exist. One—fragile X syndrome—is almost as common a cause of mental retardation as Down syndrome. It is one of the gene-expansion conditions—the mutation actually makes the gene enlarge as a result of an increased number of repeated DNA triplets. In this case, the triplet CGG is repeated over and over, as many as hundreds of times. The expanded gene, located near the tip of the long arm, causes the X chromosome that carries it to unravel or stretch out and often break. Hence, the term, fragile. Fragile X syndrome causes mental retardation in males and learning problems in some female carriers as well. There is an association between the length of the expansion and the severity of the intellectual

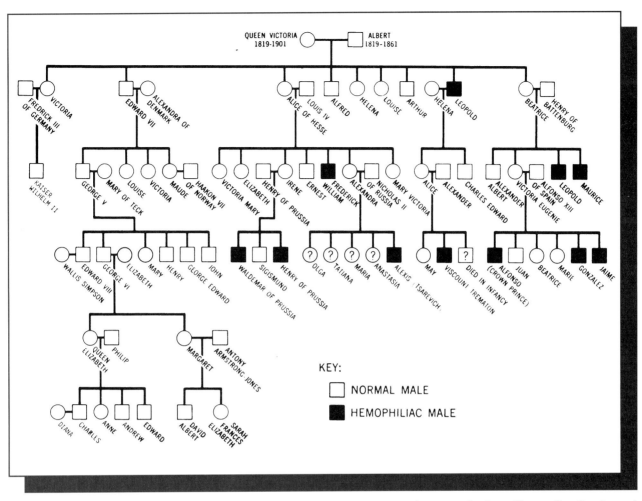

FIGURE 2 ▪ A pedigree of some of Queen Victoria's descendants, showing the hereditary distribution of hemophilia.

disability in both males and females. In addition, the condition shows anticipation (earlier age of onset and more severe symptoms) and imprinting (different expression of a gene depending on which parent passed on the mutation). Imprinting is evident in the tendency for the fragile X gene to undergo larger expansions when passed from mother to child than when passed on by the father. We look more closely at anticipation (gene expansion) and imprinting in the activities on pages 104 and 105 of this magazine.

Not all X-linked disorders are recessive. Some are dominant. For example, there is one condition in which the enamel of the teeth is very hard and its surface rough. The enamel also is very thin and cylindrical. As the enamel surface wears, the underlying dentin becomes yellow. That condition—hereditary enamel hypoplasia, or inherited, poorly developed enamel—is inherited as an X-linked, dominant trait. We see such conditions much more frequently in females.

Y-LINKED GENES

To date, we have assigned slightly more than a dozen specific genes to the Y chromosome. Ob-

viously, important male sex-determining genes, such as the gene for testis-determining factor, are on the Y chromosome, because without a Y chromosome or some crucial part of it, an individual cannot be a male. The pattern of inheritance for a gene on the Y chromosome is unique and easy to construct. Try it.

There is a small group of genes near the tip of the Y chromosome that has alleles on the short arm of the X chromosome and that does not show the typical Y-linked pattern of inheritance. That is the pseudoautosomal area of the sex chromosomes, where each gene has an allele on both the X and Y;

that is why they behave as if they were autosomal. Presumably, that phenomenon is associated with the need for the X and Y chromosomes to pair up, as do the autosomes, during male meiosis. The two sex chromosomes pair up end-to-end because of the attractive force between the same alleles present in the tips of their short arms.

HARD CHOICES FOR FAMILIES

In many case we can determine whether a fetus has an X-linked genetic disorder by prenatal tests done in the first and second trimester of pregnancy (see "Prenatal Diagnosis," page 70). Often, we can detect hemophilia, for example, by DNA analysis following amniocentesis or chorionic villus sampling. The same is true for a severe form of X-linked muscular dystrophy and a number of other disorders on the X chromosome whose genes have been cloned (genetically engineered replicas of DNA sequences) in recent years. Today, the majority of disorders

caused by mutant genes on the X chromosome have been either cloned or linked genetically to a large number of markers. The closely linked markers allow accurate identification of the X chromosome that carries a gene for a disorder within a family. We can detect individuals, usually males, who will be affected before symptoms appear after birth prenatally through cells we obtain by amniocentesis or chorionic villus sampling. Nevertheless, there is a group of serious X-linked disorders that goes undetected, the most common of which we call nonspecific, X-linked mental retardation.

For these nondetectable, serious X-linked disorders, we can determine the sex of the child prenatally through chromosomal analysis. When the doctor is reasonably sure that the mother is a carrier for a nondetectable disorder, he or she will inform the parents that there is a one-in-two chance that their male children will be affected. Female children have a one-in-two chance of being carriers, but none will have the disease.

Parents in this situation have to make hard choices. First, they must decide whether to learn the sex of the fetus by taking the test. If they choose to have the test, and the fetus turns out to be female, their worries have ended. They should, however, inform their female children that they have a fifty percent chance of carrying the gene for the disorder.

What happens if the fetus is male? The chance is fifty percent that the fetus is affected. The chance also is fifty percent that the fetus is not affected. The parents must decide whether to abort the pregnancy and risk losing a normal son or to continue the pregnancy and risk the birth of a son with the disorder.

Some people feel that abortion is wrong and that the test should not be performed in such situations. Others feel that parents should be able to decide whether to have the test and what to do once they receive the results. Difficult choices like those depend on many factors, including religious views, laws, and personal values.

DEMONSTRATING X-LINKED INHERITANCE

More than two hundred genes have been assigned to the X chromosome. They affect a wide variety of characteristics, many of which have nothing to do with sexual differentiation. Because the inheritance of these genes corresponds to the inheritance of the X chromosome, we say they are X-linked. Table 1 lists a few of the

many characteristics that are determined by genes on the X chromosome. Capital letters stand for genes with dominant effects, and small letters stand for genes with recessive effects.

To understand X-linked inheritance, remember that if a gene is carried on the X chromosome in a human male, in general, there is

no homologous gene on the smaller Y chromosome. Therefore, a trait associated with any gene on the X chromosome will be expressed, although just one gene is present. On the other hand, for a recessive trait to be expressed in a female, the specific gene must be present in both of her X chromosomes. We can demonstrate the

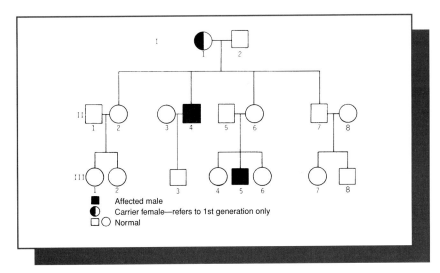

FIGURE 1 ▪ Pedigree for an X-linked recessive trait, G6PD enzyme deficiency.

inheritance of the X-linked characteristics easily using two different coins.

Materials

For each team of two students:
1 penny
1 nickel
1 copy of Worksheet 1

Procedure

Use the following example to begin the demonstration:

1. The normal gene (allele) responsible for producing antihemophilic globulin (factor VIII) is designated H. The gene for hemophilia A has a recessive effect and is designated h; it produces an abnormal factor VIII that has very little clotting activity. Suppose the father—represented by the penny—carries the normal gene H. His Y chromosome has no genes involved in blood clotting. The mother—represented by the nickel—is a carrier with one normal gene H and one recessive gene h.

2. Assign the possibilities for egg and sperm cells as follows:
   Father: penny
   Heads: X chromosome, H gene
   Tails: Y chromosome
   Mother: nickel
   Heads: X chromosome, H gene
   Tails: X chromosome, h gene
   Now, through flips of the coins, demonstrate the sex, the genotype, and the phenotype of ten offspring this couple might have. Record your findings on Worksheet 1.

3. Repeat the procedure using other characteristics and other possible genotypes in the parents. What patterns and trends do you see?

INTERPRETING THE RESULTS

1. If a man who is not a hemophiliac marries a woman who is a carrier of the gene for hemophilia, what percentage of their male offspring can we expect to be hemophiliacs?

2. The X-bearing egg and the Y-bearing sperm do not contribute equally to the genetic makeup of the embryo. Explain.

TABLE 1 X-linked characteristics.			
Characteristic	Dominant Gene	Characteristic	Recessive Gene
Normal blood clotting	H	Hemophilia A	h
Normal color vision	R	Red weakness (protanomaly)	r
Normal color vision	G	Green weakness (deuteranomaly)	g
Normal eye pigmentation	A_o	Ocular albinism	a_o
Normal G6PD enzyme	Gd	G6PD deficient	gd

3. Colorblindness and hemophilia are more common in males than in females. Explain.
4. Figure 1 shows a pedigree for the X-linked recessive trait, G6PD enzyme deficiency.
   a. What is the genotype of female II, 5? How do you know?
   b. What is the genotype of female II, 2? How do you know?
   c. What is the chance that female III, 6 is a carrier? Explain.
   d. If male II, 3 marries a female with two normal alleles for G6PD enzyme, what percentage of their sons can we expect to be G6PD deficient? What percentage of their daughters can we expect to be carriers?

X Inactivation: The Lyon Hypothesis

For many years, geneticists did not understand how a female, who was homozygous for mutant genes carried on her two X chromosomes, was no more seriously affected than a male, who had only one such gene on his one X chromosome. Geneticists wondered why two "doses" of the gene seemed to have no greater effect than one.

In 1961, several groups of scientists proposed a hypothesis to answer those questions. The first person to explain the hypothesis precisely and in detail was Mary F. Lyon, a British geneticist. The hypothesis was originally called the Lyon hypothesis in her honor.

The phenomenon that Mary Lyon explained is called X inactivation. It has four main parts: (1) one of the X chromosomes is inactivated in each of the somatic (body or nonreproductive) cells of a normal female; (2) the inactivation is random—either the X from the mother or the X from the father may be inactive in any single cell; (3) once inactivation occurs in a cell in a developing embryo, the same X chromosome will be inactivated in all the cells that descend from that cell; and (4) the inacti-

vation occurs early in the development of the embryo, at about sixteen days after conception.

Lyon began to formulate the idea of X inactivation when she observed female mice that were heterozygous for X-linked genes for coat color. Those females had patchwork fur. Males did not show a patchwork coat color, and neither did mice that had only a single X chromosome (the XO chromosomal makeup).

In 1949, Murray Barr and Ewart Bertram began to study a dark-

stained mass that appears in the nucleus of certain cells. Although scientists noted those bodies as early as 1909, Barr and Bertram were the first to show that they normally are present only in female cells. In humans, we find that these Barr bodies adhere to the inner wall of the nuclear membrane (Figure 1). Susumo Ohno, an American geneticist, suggested that the Barr bodies might be made up of materials from sex chromosomes. Recalling her mouse studies, Mary Lyon hy-

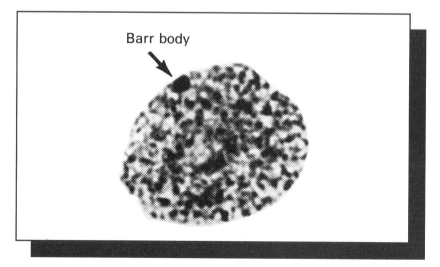

FIGURE 1 ▪ A cell with one Barr body.

pothesized that the Barr body is an inactivated X chromosome. That would explain the difference in "doses" between males, who have one X chromosome, and females, who have two X chromosomes.

The most convincing evidence in support of the Lyon hypothesis came from studies of an enzyme called G6PD. In humans, that enzyme is produced by an X-linked gene. Several investigators showed that there is no difference in the activity of the enzyme between normal males and females. And, there are no differences among individuals with three or more X chromosomes. Thus, the dose—the number of genes present—does not seem to make any difference in the amount of the enzyme that is present or in the action of the enzyme.

Other studies were done on cells from heterozygous females who had two different alleles for different forms of G6PD on each X chromosome. In other words, one allele on one X chromosome coded for one form of the enzyme and the other allele on the other X chromosome coded for a different form. Biochemical tests can tell the difference in the enzymes that are produced. When scientists grew cells from these heterozygous females in cell culture, they found that only one form of the enzyme was produced in the progeny of any single cell. Thus, it appeared that only one X chromosome was active, the other being inactivated. Those results also suggested that the same X is inactivated in all the cells that descend from any single cell.

As far as we know, X inactivation occurs only when there are at least two X chromosomes present. It does not occur in 46, XY or 45, X individuals. And, all but one X chromosome will be inactivated in a cell. For example, males with Klinefelter syndrome—47, XXY—will have one Barr body in each cell. Females who are 47, XXX will have two Barr bodies in each cell. How many Barr bodies would you find in the cells of an individual who is 48, XXXX?

X inactivation helps us understand why X-linked conditions tend to have a highly variable expression in female carriers. Female carriers of hemophilia may be as severely affected as males, not affected at all, or have bleeding tendencies somewhere in between. Similarly, a small percentage of female carriers of muscular dystrophy have varying degrees of muscle weakness. In theory, the variability among carriers for X-linked alleles results from the random inactivation of X chromosomes. Inactivation early in the embryonic development of most of the X chromosomes that carry the gene for hemophilia or muscular dystrophy (that is referred to as skewed X inactivation—a major deviation from the expected fifty percent) can lead to a female who exhibits either mild or no symptoms. Conversely, inactivation of most of the normal alleles can lead to the severe symptoms that we sometimes observe. Recently, using newly developed DNA procedures, biologists have shown that all of the symptomatic carriers of the X-linked gene for muscular dystrophy have skewing of X inactivation, thus confirming the theory.

X inactivation also accounts for the relatively mild clinical effects of having one or more extra X chromosomes. Individuals who have an extra X chromosome may have few, if any, medical problems (see "The Boy with the Extra X Chromosome," page 88). The presence of an extra autosome, as with Down syndrome and other trisomies, produces much more serious effects.

Important questions about X inactivation remain, although much research has been done and is continuing. For example, is the entire X chromosome inactivated? Clearly, the answer is no. We know that about a dozen X-linked genes escape the process; most of those genes are clustered near the tip of the short arm of the X, the area of homology (identical with respect to genes present) with the tip of the short arm of the Y. Those parts of the X and Y are pseudo-autosomal, because, like the autosomal chromosome pairs, they contain matching alleles. Presumably, the escape from inactivation has something to do with the necessity of the X and Y to pair end-to-end during male meiosis.

Another intriguing mystery, well on the way to solution, is the mechanism responsible for inactivation. Is it really a random process? Does it remain forever "imprinted" on the inactivated X? How does it come about?

Many years ago, geneticists observed that in certain animals the paternal X is always inactivated in somatic cells. More recently, experiments in both mouse and human tissues have shown that in the extraembryonic parts (e.g., the membranes forming the amniotic sac), the paternally derived X chromosome is preferentially inactivated. In addition, if you think about the production of mature ova in the female, it becomes obvious that X inactivation must be reversible because every female randomly passes one or the other of her X chromosomes to her children via her ova and whichever X gets there, it has to be active. Thus, X chromosome inactivation is not random for the extraembryonic structures; inactivation must become uncoupled in meiosis of females and must reoccur randomly in the embryo itself.

Another missing piece of the puzzle is the mechanism for selecting which X is to become inactivated in the female embryo at about sixteen days after conception. In the mouse, biologists have found a gene locus that contains the controlling element for X inactivation. Apparently, the product(s) of this gene randomly selects X chromosomes for inactivation in the embryo and some sort of interaction between other genes maintains it. Mary Lyon has continued to play a major role in the research that has led to the new data on the biochemistry and molecular biology of gene action in the X chromosome.

Here are some data resulting from the observation of natural events. On a separate sheet of paper, indicate whether they are consistent with or in conflict with X inactivation. You should explain each of your answers. The following information will help:

- The genes for fur color in mice and cats are carried on the X chromosome.
- Barr bodies are dark-stained masses in the nucleus of certain cells of XX females, as shown in Figure 1.
- Hemophilia is a recessive, X-linked trait that produces severe symptoms in males.

Observations

1. Female mice that were heterozygous for X-linked genes for coat color have fur that is made up of patches of the different color.
2. Male mice do not exhibit the patchy coat color seen in females.
3. Female mice with a single X chromosome (XO) do not show the patches of different color seen in XX females.
4. A Barr body appears to be a condensed X-chromosome.
5. The enzyme G6PD is coded for by an X-linked gene in humans. There is no difference between normal females and males in the amount of enzyme produced.
6. The amount of G6PD in individuals with three or more X chromosomes is the same as that found in XX females.
7. The gene that codes for the production of the steroid sulfatase enzyme is X-linked. Females normally produce twice as much of that enzyme as do males.
8. There are two forms of G6PD, each coded for a different X-linked allele. Individual cells from females heterozygous for G6PD produce only one form of the enzyme.
9. Males with Klinefelter syndrome (XXY) have one Barr body in each cell.
10. Female carriers of hemophilia may be as severely affected as males, not affected at all, or have symptoms somewhere in between.
11. In kangaroos, the inactivated X chromosome is always the paternal X.
12. Calico cats, nearly all of which are females, are heterozygous for a coat-color gene (black and yellow). The rare male cat that appears as calico (black and yellow patches of fur) turns out to be an XXY male.

SEX AND GENES

Study the pedigree in Figure 1 for the common male pattern baldness. What do you think is the type of inheritance?

Only males are shown to be affected. We hope that you didn't think for a moment that this is X linkage. Why not?

What we are seeing here is autosomal dominant inheritance that is sex-influenced (the trait is modified by the sex of the individual) or sex-limited (the trait is completely suppressed in one sex or the other). The gene is probably present in some of the apparently unaffected females in the family, but premenopausal females are protected from its effects, most likely by estrogen and possibly other sex hormones. The internal environment, as well as the external environment, of our genome can influence the expression of

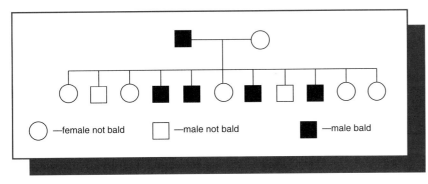

FIGURE 1 ▪ Pedigree for pattern baldness.

female not bald ⬤ **male not bald** ⬛ **male bald**

genes. Presumably, male sex hormones, including testosterone, play a role in triggering the hair loss in males. Some females who inherit the gene for male pattern baldness do have significant hair loss after menopause when the levels of female sex hormones decrease. Thus, the condition is not totally sex-limited, but certainly sex-influenced.

There are many other sex-influenced or sex-limited traits. A woman does not normally have a beard, yet she carries all the genes necessary to produce one. The beard type of her sons depends as much on genes they receive from their mother as those from their father. On rare occasions, the genes for sex-influenced or sex-limited traits are expressed in the wrong sex; that is, for some reason, the genes escape the normal controls on expression. For example, some women have minor beard growth and occasionally accompanying excess body hair. Additional obvious examples of sex-limited characteristics include the normal development of breasts and differing genitalia in females as compared to males.

Keep in mind that we cannot account for all sex differences by this mechanism. For example, in the earlier decades of the twentieth century when society was quick to institutionalize mentally disabled people, there were significantly more male occupants of these institutions than female. We assumed that mentally disabled males, on the average, were bigger, stronger, and more aggressive than females. Thus, they would be more likely to be a problem to their families and communities and proportionally more were institutionalized—a kind of sex-limited phenomenon. That has turned out to be only part of the explanation. We now know that there are many genes on the X chromosome that cause mental disability, with or without characteristic physical features, and those account for a large part of the excess of males with mental handicap.

A puzzling group of birth defects, exemplified by pyloric stenosis, shows remarkable sex differences. Pyloric stenosis is an overgrowth of the muscle that controls the flow of stomach contents to the rest of the intestine at the junction of the stomach and the small bowel—the pylorus. The overgrowth narrows the opening and the result is vomiting and eventually nearly total obstruc-tion. Many of these infants used to die of severe dehydration and starvation.

Pyloric stenosis usually occurs in infants between four to six weeks of age and is easily treated by minor surgery. It is a fairly common condition and the cause is polygenic (several poorly functioning genes are responsible). As expected for polygenic and multifactorial conditions, the siblings and offspring of an affected individual develop pyloric stenosis far more frequently than we can account for by chance.

The extraordinary finding is the sex ratio: males are affected nearly five times more often than females; that is, the incidence is about 1 in 200 male infants, compared to about 1 in 1000 females. The theory is that there is a threshold effect; for females to develop pyloric stenosis, they must inherit either a larger number of the causative genes or a more potent set of them. Whatever the explanation, it protects female infants from the effects of the same group of genes that gets males into trouble. That could only be the result of the different internal environments between the sexes even as early as infancy—the differing proportions of sex hormones and possibly other gene products that we do not know about.

Can you can figure out the relative risks for offspring of individuals affected with pyloric stenosis based on the sex of the affected parent? Would a male who had pyloric stenosis be more or less likely to have affected offspring than a female? Why? Would an affected female parent be more likely to have affected daughters than an affected male parent? Why?

IMPRINTING

The term imprinting probably was first used in biology in the 1930s to describe modification of all kinds of behavior because of particular experiences. For example, we've seen newly hatched goslings "imprinted" to behave as if a dog were their mother if a dog was the first moving object they saw after hatching.

Chromosomal imprinting describes the selective inactivation of paternally derived X chromosomes in extra embryonic membranes, as described in "X Inactivation," page 100. It also occurs in other chromosomes or segments of chromosomes (see following discussion). We don't know what mechanism is responsible for that type of imprinting or why it is the paternal rather than the maternal X that is selected. In this article, we will look at genomic imprinting, a different expression of genetic material, either at a chromosomal or single-gene level, depending on whether the genetic material has come from the mother or father.

Let's look at an example in Figure 1. A mutation (a small deletion, in this case) in human chromosome 15—assuming chromosome 15 came from the father—causes Prader-Willi syndrome (PWS). The main symptoms of the disorder are mental disability, obesity due to an uncontrollable appetite, and a characteristic facial appearance. The affected child may be male or female. The father does not have PWS or the deletion. PWS nearly always occurs as a new mutation, in this case arising in the individual sperm from which the affected child develops. It is the parental origin of the chromosome that is important.

The identical mutation, if inherited from the mother, causes a completely different syndrome called Angelman syndrome (AS). The main symptoms here are a very different facial appearance; an abnormal, stiff gait or way of walking (like a puppet); and mental disability with convulsions and inappropriate bursts of laughter. Again, the child may be of either sex and the mother is normal. The new mutation this time arises in the ovum.

Genomic imprinting must involve some modification of the nuclear DNA in somatic cells to account for these extraordinary differences between PWS and AS. That is very much contrary to Mendel's laws, which state that genes pass unchanged from parent to offspring and the parent of origin makes no difference. Furthermore, genomic imprinting means that something happens during a critical period of development, possibly when the cells that will give rise to the ova or sperm (the germ line cells) are forming in the embryo.

Uniparental disomy (UPD, literally, two chromosomes from the same parent) is another form of genomic imprinting. Using PWS and AS as examples again, some individuals with those conditions have no deletion at all. Instead, in

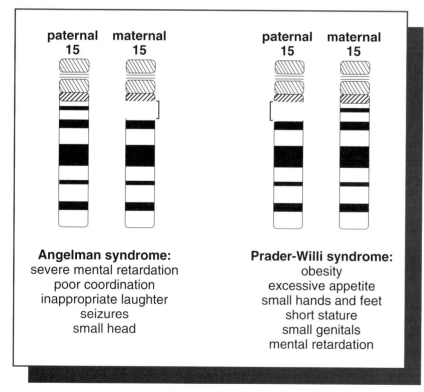

Angelman syndrome:
severe mental retardation
poor coordination
inappropriate laughter
seizures
small head

Prader-Willi syndrome:
obesity
excessive appetite
small hands and feet
short stature
small genitals
mental retardation

FIGURE 1 ■ Genomic imprinting results in different phenotypic express in these two human genetic disorders.

PWS, for example, the child may have two copies of the mother's chromosome 15 and no chromosome 15 from the father. How could that happen? Suppose the conception started as trisomy 15—three chromosome 15s—two of which came from the ovum as a result of a maternal nondisjunction. (That is the same mechanism for Down syndrome's trisomy 21; see "Robert Vandenberg Wins Special Olympics," page 53.) In the first cell division of the embryo, an "error" could occur and one of the three chromosome 15s is lost. If it were the one from the father, the embryo would have only the two maternally derived 15s. The total number of chromosomes would be forty-six, down from forty-seven, and you would think the child ought to be normal. But, because of some type of genomic imprinting, humans require at least part of a paternal 15 for normal development. Having two maternal 15s is the same as having a paternal deletion in the PWS critical area, and hence, the child has PWS. The reverse chromosomal situation, UPD for the paternal chromosome 15s, accounts for a small percentage of cases of Angelman syndrome.

Recent studies have identified several other syndromes where the parental origin of a deletion, or UPD involving chromosomes or segments of chromosomes other than 15, influences or modifies the symptoms.

These new twists on Mendelian genetics illustrate the nature of science. Scientists expect that new information and new observations will cause them to modify their explanations for natural phenomena. Science is an open-ended inquiry into the natural world, and scientists must always be open to new ideas based on sound scientific methods.

Our understanding of imprinting results from our ability to conduct experiments and make observations that were not possible in the mid-1800s, when Mendel did his pioneering work. The modification of Mendel's explanations of heredity does not mean that his explanations were wrong—only incomplete.

ANTICIPATION: SOME GENES DO CHANGE

One of Mendel's most important observations was that his postulated hereditary factors, now obviously the genes, are passed unchanged from generation to generation. Genetic anticipation is a startling exception to that fundamental law, so startling, in fact, that few geneticists believed it to be possible. The term refers to the apparent worsening of the effects of a gene as it passes from grandparent to parent to child to grandchild, and so on. In each generation, the disorder has a progressively earlier age of onset and more severe symptoms.

The resistance to accepting that concept, in addition to its flying in the face of Mendel's laws, was strengthened by the political and social times in which it was first proposed. In the 1920s and 30s, several groups of geneticists attempted to demonstrate that when mentally retarded parents had children, those children were likely to be more retarded than their parents. The Nazis eagerly adopted those notions to support their campaign to eliminate "inferiors" by sterilizing the retarded. Sad to say, there were geneticists in America who also believed that sterilization of the incompetent was essential for the preservation of intellect in the human race. The biases in the family studies and the flaws in the conclusions were documented particularly well by Professor Lionel Penrose of the Galton Institute of Human Genetics in London, England.

It also was in the 1930s that one of Penrose's colleagues at the Galton Institute studied families with myotonic dystrophy (MyoD) and collected and published pedigree evidence for anticipation. MyoD is an autosomal dominant condition that causes muscles to fail to relax after contracting. The patient has difficulty letting go of things.

Over time, varying from months to many years, the myotonia, or difficulty with muscle relaxation, is replaced by generalized muscle weakness not unlike what occurs in muscular dystrophy. In addition, the MyoD gene causes cataracts (clouding of the lens of the eyes), early onset of male pattern baldness, occasional irregularities of the rhythm of the heart, and in some fifteen percent to twenty percent of individuals, learning problems of varying severity. The pedigrees, especially when carefully researched, usually reveal individuals in early generations of the family with

cataracts, but no detectable signs of the muscle problems. In the next generation, there would usually be individuals with relatively mild muscle problems with or without cataracts, and perhaps some of the other problems. In the next generation, there were even more severe symptoms.

Professor Penrose himself, under the influence of his battles to disprove anticipation for mental retardation, dismissed the MyoD

data as statistically biased. What arguments might the professor have used to counter the apparent anticipation in the MyoD pedigrees?

It took until the 1990s for the explanation of anticipation to emerge with the demonstration that some genes do change as they pass from generation to generation and that those changes cause the earlier ages of onset and increasing severity of the symptoms. The phenomenon is allelic expansion and the mechanism is trinucleotide repeats—a mutation where three basepair repeats (the basepairs are the components of DNA and RNA, see the appendix "The DNA Molecule," page 139) occur in tandem (one after another), thus enlarging the gene and causing instability. The gene actually gets bigger and eventually its function is disrupted to varying degrees. In some gene-expansion diseases, such as MyoD (Figure 1) and the fragile X syndrome, the expansion increases from generation to generation and results in increasing severity. In Huntington disease, however, the expansion is quite small and remains fairly stable as it passes down through a family (see Figure 2).

Another curious phenomenon in some of these gene-expansion diseases is the variation in severity based on the sex of the parent who transmits the gene. In MyoD families, sometimes affected infants are born with severe muscle weakness, in contrast to the usual age of onset, which is the early teenage years. Those infants are affected only when the mother has the disease, never when the gene is passed from the father. Scien-

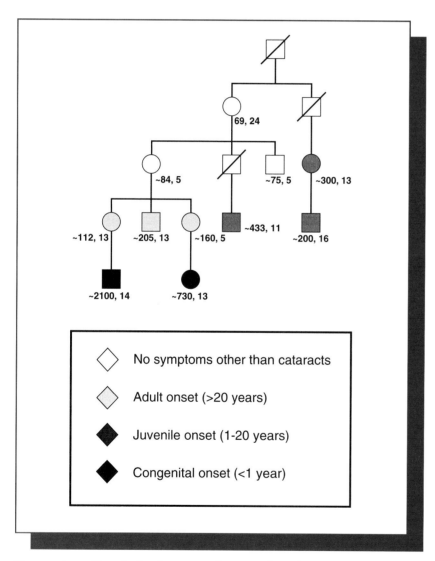

FIGURE 1 ■ Correlation between the trinucleotide repeat number and the clinical features of myotonic dystrophy. The numbers represent how many copies of the trinucleotide repeats are present in each allele.

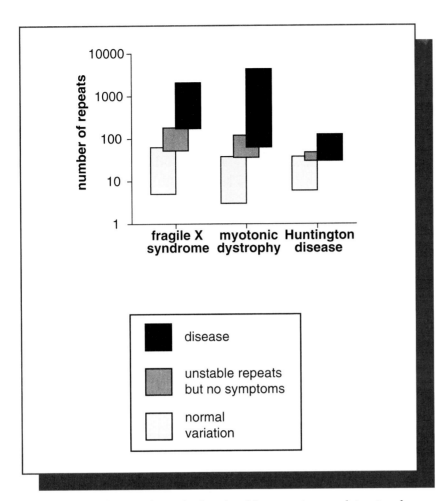

FIGURE 2 ▪ The number of trinucleotide repeats correlates to phenotype for certain genetic disorders. Notice the overlap of symbols, suggesting that the number of repeats is not the only factor to influence manifestation of the disorder.

tists thought that phenomenon might be another example of imprinting (see "Imprinting," page 78), but it probably results from the tendency of the gene to undergo greater expansion when transmitted from females than males. We don't know how that happens and, in fact, one day scientists may show that indeed it is some form of imprinting. Curiously, in Huntington disease, one of the small-expansion conditions, the rare childhood onset occurs only when the affected child receives the gene from the father.

The number of human diseases that we know result from gene expansion is small (fewer than ten at present), and all affect the brain and/or nervous system. We don't know why allelic expansion occurs.

MITOCHONDRIAL INHERITANCE

Study the pedigrees in Figure 1. The affected individuals (black symbols) have hearing loss, mild muscle weakness, and learning problems. Could any type of Mendelian heredity explain the pattern of affected family members? Think carefully before you read on.

At first glance, it might look like an autosomal dominant disorder, because there are affected individuals in each generation and males and females seem equally likely to be affected. But, something's wrong. In generation II, *all* of the offspring of the affected female in generation I are affected. With a fifty-fifty chance, that would be possible, but certainly unusual. What is the chance of having six affected and no unaffected children if it were an autosomal dominant disorder?

Could it be an X-linked dominant trait? If not, why not?

What we are seeing in Figure 1 is a typical pedigree for mitochondrial or maternal inheritance. All of the children of an affected mother will be affected and none of the children of an affected father.

The mitochondria are major energy-producing factories in most cells of higher organisms.

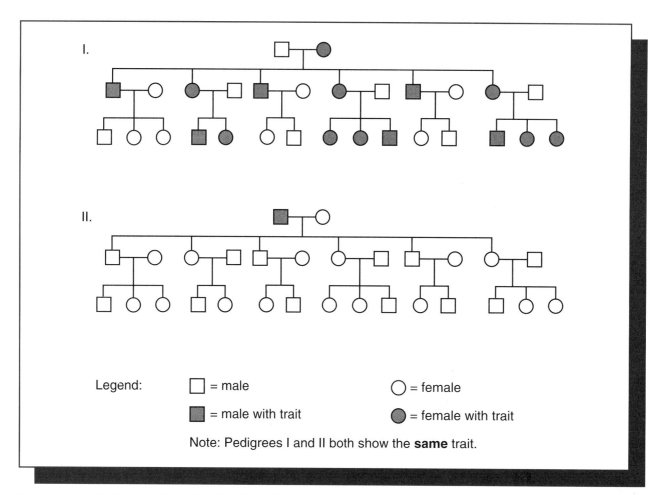

FIGURE 1 ■ Pedigrees of two families that display the same trait.

They use molecular oxygen and a series of respiratory enzymes to produce energy rapidly. It is not surprising, then, to learn that mitochondrial mutations are likely to affect muscles and the brain, organs that require large amounts of energy, often on short notice. Mitochondria contain their own DNA in the form of a ring. That DNA codes for some of the enzymes present in the mitochondria, as well as for proteins that make its own physical structure. Other enzymes present in mitochondria are "imported" from the cytoplasm after having been coded for by nuclear genes.

To a considerable extent, the structure and function of mitochondria, including the circular configuration of the DNA, are reminiscent of bacteria. It is quite likely that the mitochondria of higher organisms evolved from bacteria that invaded early life forms and then became incorporated into their structure. Mitochondria are not present in organisms such as viruses, bacteria themselves, or yeast; they make their first appearance in the evolution of life in the algae.

Why do we see this peculiar type of pedigree when the cause of the disorder is a mitochondrial mutation? That's an easy question to answer. The ovum contains all of the mitochondria that we inherit. They are present in the cytoplasm of the egg and after fertilization, they replicate along with the nuclear DNA with each cell division of the embryo.

But, don't the sperm have mitochondria to provide the energy to keep those tails wagging to propel them on their way? Sure they do, but where do they appear in the sperm structure and what happens to them when fertilization is achieved?

It may seem strange to put the genetics of cancer in a section on exceptions to Mendel's laws, but the reasons should become clear as you read about breast and ovarian cancer, retinoblastoma, and lung cancer. We'll begin with a couple of examples that seem, at first look, to fit nicely into predictable Mendelian pedigrees. Keep in mind as you work through these examples that every human possesses dozens, if not hundreds of genes that function poorly; that is, they produce nothing at all or a product that is so different from the usual that the function is impaired (see "Why Study Human Genetics?," page 4). Most of the time, those poorly functioning genes do no harm because our genes come in pairs and the "normal" allele will provide enough normal product for normal function. Sometimes, however, unique combinations of these poorly functioning genes, or even single ones, can cause or predispose an individual to birth defects, cancer, heart disease, or other health risks. Several genetic factors can act together—the condition then has a polygenic cause—and sometimes in conjunction with environmental factors—the multifactorial conditions.

BREAST AND OVARIAN CANCER

At present, we don't know the relative extent to which genes and environment contribute to the cause of breast and ovarian cancer, nor to the majority of other forms of cancer, for that matter. However, we do know that:

- the risk of developing breast and ovarian cancer is increased among females whose close relatives have developed breast and/or ovarian cancer,
- the risk may be increased further if the breast cancer occurred in relatives before menopause, and
- between five percent and fifteen percent of females with breast cancer have a significant family history of the disease. (These are the so-called inherited forms of breast cancer.)

Susceptibility Genes for Breast Cancer

BRCA1 (breast cancer 1) and *BRCA2* (breast cancer 2) are genes with a dominant effect that put members of a family at an increased risk for developing breast, ovarian, and other cancers. The mutations in those genes that predispose to cancer probably act by disrupting their function as tumor suppressor genes, a concept discussed in the article "Retinoblastoma," page 113.

BRCA1, located on the long arm of chromosome 17, was the first to be isolated and accounts for about fifty percent of all *inherited* breast cancer, or about five percent of *all* breast cancer. About 400,000 women in the U.S. have one of the more than one hundred mutations already detected in the *BRCA1* gene. Some of the changes are apparently harmless.

We think that *BRCA2*, located on the long arm of chromosome 13, accounts for thirty percent to forty percent of all inherited breast cancer. Undoubtedly, other breast cancer genes exist, and scientists will isolate them in the future.

General Population Risks for Developing Breast Cancer

The risk for developing breast cancer in the U.S. population in general is about eleven percent by age eighty-five. The lifetime population risk for ovarian cancer is about one percent. Thus, cancer screening is important for all women; women should determine the frequency of physical examinations by their age and by the results of ongoing research.

If a woman inherits an altered *BRCA1* or *BRCA2* gene, her lifetime risk for developing breast, ovarian, and possibly other cancers is increased beyond the risk for the general population. **Not all women who inherit an altered breast cancer gene, however, will develop breast, ovarian, or any other cancers.**

Cancer Risks Associated with BRCA1

Breast cancer

- The risk is low before age thirty (although we have seen rare cases in women as young as nineteen).
- From thirty to forty-five, the risk rises rapidly to fifty percent.
- From forty-five to eighty-five, the risk rises more slowly to eighty-five percent and ninety percent.

Ovarian cancer

- The lifetime risk may be as high as sixty percent and seventy percent.
- The risk rises gradually from the thirties, reaching forty percent by age seventy.

Colon and prostate cancer

- The risk of colon cancer in men or women with an altered *BRCA1* is slightly increased over the general population risk.
- Similarly, the risk for men with an altered *BRCA1* is slightly increased.

In neither case is the risk believed to be sufficiently increased to institute any special screening or other forms of examination.

Cancer Risks Associated with BRCA2

- *BRCA2* mutations affect cancer risks for both males and females.
- Risks for breast cancer in females are increased about the same as for *BRCA1*, but not as high as *BRCA1* for ovarian cancer (fifteen percent to twenty percent by age seventy).
- The risk of male breast cancer, a rare cancer in the general population (lifetime risk of about 0.1 percent), rises to about six percent by age seventy.

And Now, the Problems: BRCA1 *and* BRCA2 *Testing Is Highly Controversial*

The test can detect mutated *BRCA1* and *BRCA2* genes with ninety percent to ninety-five percent accuracy. Current technology has limited sensitivity and will miss some mutations. Once a mutation is identified within a family, however, the testing of relatives at risk is close to one hundred percent accurate.

Who gets tested? Should we screen all of the roughly 400,000 women in the U.S. who might have mutations in the *BRCA1* gene? Not all familial breast or ovarian cancer is associated with altered *BRCA1* and *BRCA2*. To start testing in a family, one affected member must agree to be tested for *BRCA1* and *BRCA2* and must be willing to learn the results. Then other family members can be tested, if they so wish.

What if you test positive? Each family member must develop a personal risk-management program. At present, the options for breast or ovarian cancer include:

- breast self-examination (the recommendation is once a month);
- regular examinations by a physician, every six months;
- mammogram and pelvic ultrasound of the pelvis every six months;
- surgery; some physicians recommend the early removal of breast tissue (mastectomy) and ovaries. Obviously, that is a radical approach, and unfortunately, it doesn't remove the risk totally for either type of cancer. Breast cancer can occur after a mastectomy and ovarian cancer after removal of the ovaries. Apparently, some breast and ovarian tissue can remain no matter how skilled the surgeon.
- medication/diet; so far, there is no evidence that chemotherapy, vitamins, special diets, or anything else decreases the risk for those with altered *BRCA1* or *BRCA2*. We are studying some of the cancer treatment drugs to find out whether they might have preventive effects.
- risk avoidance or avoiding environmental exposures that may increase the risk of breast, ovarian, colon, prostate, or other cancers. Unfortunately, unlike lung cancer and smoking, we don't know anything about the interactions between susceptibility to genes for breast cancer and environmental factors that might increase or decrease the risk.

Now you can understand why *BRCA* mutation screening is controversial. If you had more than one close relative with breast, ovarian, colon, or prostate cancer, would you be screened for *BRCA1* and *BRCA2*? Why? Some constituencies are advocating the availability of *BRCA1* and *BRCA2* screening for the entire female population. What do you think of that?

You can learn more about *BRCA1* and *BRCA2*—including screening and testing—by consulting the Web site for the National Action Plan on Breast Cancer: http://www.napbc.org. We also have included a brochure titled "Understanding Genetic Predisposition to Breast and Ovarian Cancer," pages 111–112.

Understanding Genetic Predisposition to Breast and Ovarian Cancer

It is well known that breast and ovarian cancer often occur in women whose relatives have had these diseases. We now know that this may be due to the inheritance of a mutation—*an alteration in the DNA sequence— in either of two susceptibility genes called BRCA1 and BRCA2.*

This pamphlet describes a test for BRCA1 and BRCA2 mutations which is now available and may be important to your health care if you have a personal or family history of breast or ovarian cancer. Knowing about your cancer risk may help you and your doctor make better-informed decisions both for early detection and treatment of these cancers. After reading this pamphlet and evaluating your personal risk, you may wish to discuss genetic testing further with your doctor.

UNDERSTANDING HEREDITARY RISK

Overall, American women have a 1-in-8 chance of developing breast cancer and a 1-in-70 chance of developing ovarian cancer during their lifetime. For some women, however, the risk is much higher. Women at increased risk of breast and ovarian cancer generally belong to families in which other members of the family have had these diseases. In some families, cancer is caused by a gene mutation that is passed down from mothers or fathers to their children.

Everyone is born with two copies of about 100,000 different genes. One copy of each gene comes from your mother, the other from your father. A gene called BRCA1 was the first gene found to play a major role in hereditary breast and ovarian cancer. A second gene, BRCA2, can also increase the risk of hereditary breast and ovarian cancer. Normally, these genes help to *prevent* cancer by making proteins that keep cells from growing abnormally. However, if a woman inherits a mutation in BRCA1 or BRCA2 from her mother or father, this protection is lost and her risk of breast or ovarian cancer is increased.

THE TESTING DECISION

Genetic testing for susceptibility to breast and ovarian cancer is not for everyone. It's not like a mammogram or other screening test intended for the general population, but rather is intended for individuals thought to be at high risk of breast and ovarian cancer because of their family history, or because they themselves developed breast cancer at an early age. The questionnaire included in this pamphlet is about personal and family characteristics associated with the risk of hereditary breast and ovarian cancer. If you answer "yes" to two or more of these questions, you may want to ask your doctor or a genetic counseling professional for more information about genetic susceptibility testing.

Before deciding to have genetic testing, it's important for an individual to fully discuss the analysis and its possible implications with his or her doctor or genetic counseling professional. Every individual who decides to be tested is asked to sign an informed consent form, which is an agreement between the individual and the doctor showing that they have discussed the test and its possible implications for the individual and his or her family.

■ *By age 70, up to 87% of women with a BRCA1 mutation develop breast cancer, and up to 44% develop ovarian cancer. For women who have BRCA2 mutations, the risk of breast cancer is increased to roughly the same degree as it is with BRCA1 mutations. Women with BRCA2 mutations also may face an increased risk of ovarian cancer. (Keep in mind that these risk figures are calculated from studies in high-risk families, and individual risk may vary.)*

■ *Breast cancers that are caused by BRCA1 or BRCA2 mutations usually occur relatively early in life. For example, in women with BRCA1 mutations, the average age at diagnosis is 45 years, compared to 64 years for breast cancer not related to BRCA1 mutations.*

■ *Certain mutations in the BRCA1 and BRCA2 genes are especially common among Ashkenazi Jews, whose ancestors come from central or eastern Europe (most American Jews are of Ashkenazi descent).*

■ *Although breast cancer is rare in men, a man may carry BRCA1 or BRCA2 mutations, which can be passed on to his children. For example, a man who has a mother or sister with a known BRCA1 or BRCA2 mutation may also have inherited the mutation. If he did, each of his children has a 50/50 chance of inheriting the mutation from him.*

Genetic testing has important implications whether the results are positive, negative, or of uncertain significance. A positive result means that the test found a mutation that may increase the individual's risk of breast and ovarian cancer. A negative result means that no mutation was found. A result of uncertain significance means that a mutation was found but it is not yet known whether the mutation increases the risk of cancer.

For many people, knowing their test results is important, because this information may help to guide health care decisions for themselves and their families. Before deciding to be tested, however, it's important for individuals to think about how they and their families might deal with the issues raised by knowing their test results. If you are thinking about genetic testing, ask your doctor or genetic counseling professional for more information.

WHAT IS BRACANALYSIS™?

Scientists at Myriad Genetics led the teams that discovered the complete sequences of the BRCA1 and BRCA2 genes, and subsequently Myriad Genetic Laboratories developed a test, called BRACAnalysis™, to identify mutations (sequence alterations) in these genes. BRACAnalysis™ is a highly technical process for analyzing the full sequence—each chemical building block—of the DNA (genetic material) in an individual's blood sample. DNA sequencing—the method used in BRACAnalysis™—is considered the "gold standard" of genetic testing because it identifies mutations not found by other methods.

If you would like more information about BRACAnalysis™, call Myriad Genetic Laboratories at 1-800-469-7423.

MYRIAD

Myriad Genetic Laboratories, Inc.
320 Wakara Way
Salt Lake City, Utah 84108-9930

For more information:

Toll free (800) 469-7423
FAX (801) 584-3615
e-mail BRACA@myriad.com
website http://www.myriad.com

BRCA1 and BRCA2 genetic susceptibility analysis may be considered investigational in some states.

MY6802(200)   January 1997   Printed in the USA

ANSWER THE QUESTIONS BELOW TO EVALUATE WHETHER OR NOT YOU MAY BE A CANDIDATE FOR GENETIC SUSCEPTIBILITY TESTING FOR BREAST AND OVARIAN CANCER.

1. Have you been diagnosed with breast cancer before age 50?

☐ *Yes*   ☐ *No*

2. Have you been diagnosed with ovarian cancer (at any age)?

☐ *Yes*   ☐ *No*

3. Do you have relatives who have been diagnosed with breast cancer before age 50?

☐ *Yes*   ☐ *No*

4. Do you have relatives who have been diagnosed with ovarian cancer?

☐ *Yes*   ☐ *No*

If you answered "yes" to two or more of these questions, you may want to ask your physician or a genetic counseling professional for more information about genetic susceptibility testing and BRACAnalysis™.

RETINOBLASTOMA

In addition to cancers as common as breast and ovarian cancer, some rare forms of cancer are inherited in a pattern that fits with Mendel's laws. One well-studied example is retinoblastoma—a type of cancer that affects the retina of the eye. Retinoblastoma occurs in infancy or early childhood and can be treated by removing the affected eye or eyes, or in some cases by radiation or laser treatment to the area. Although many children will lose their vision in the affected eye or eyes, early detection and treatment may result in preservation of sight and a normal life span. This cancer is inherited as an autosomal dominant trait in some families, and research on the disorder has taught us much about the interrelationships between genetics and cancer.

In the families where retinoblastoma is inherited, the tumors almost always occur in both eyes. In the cases where retinoblastoma happens only once in a family (sporadic cases), however, only one eye is involved most of the time. Those findings led Dr. Alfred Knudson, at the Institute for Cancer Research, Fox Chase Cancer Center, Philadelphia, to suggest that two events, "the two-hit theory," are necessary for retinoblastoma to arise.

For the inherited cases, the first event is a gene mutation that already is present in the retinal tissues because it is inherited from one of the infant's parents. The parent either will be affected (he or she survived because of successful treatment) or will have passed on a new mutation that

occurred in the ovum or sperm from which the affected infant developed. The second hit, a chance event, occurs in the normal allele of a single retinal cell, and the removal of its suppressive effect allows the retinoblastoma gene to do its tumor-producing work. In other words, it takes a hereditary mutation plus a somatic mutation to cause the cancer.

In the sporadic cases, both events must occur in the developing retinal tissue. What that means is that one allele at the retinoblastoma gene locus in a single retinal cell mutates to the retinoblastoma gene. Then the other allele, probably at a different time, undergoes a mutation that cripples its suppressor effect. Then a tumor develops.

Now, pause for a moment, and see whether you can figure out the answer to these two questions.

1. Would you expect the inherited or the sporadic type of retinoblastoma to show up earlier (at a younger age, on the average) in an affected infant? Why?
2. Would you expect the inherited or the sporadic type to be more likely to produce multiple tumors in the same eye? Why?

If you picked the inherited type as the answer for both, you are correct. The inherited tumors start earlier because the predisposing mutation was present from the time of conception and is in every retinal cell, susceptible to that second hit. In fact, the tumor or tumors may be present and easily

detectable at birth, whereas the sporadic retinoblastoma may not appear until late infancy or even into early childhood. The reason for the latter should be evident—two separate mutations have to occur in the retinal cell to give rise to the tumor. That is, both alleles at the gene locus associated with retinoblastoma have to mutate, and that takes more time. For the same reason, the inherited form is much more likely to produce multiple tumors than the sporadic form; again, all of the retinal cells start out with a first hit.

Chromosome analysis in some cases of retinoblastoma also provided further understanding. Some children who had retinoblastoma also had deletions (missing regions) for a specific part of chromosome 13. That led scientists to propose that the gene responsible for retinoblastoma is localized to that region. Further studies involving gene mapping and cloning proved that to be true, and in 1987, scientists identified the gene for retinoblastoma. Finding the gene has led to improvements in diagnosing individuals who carry the gene. We now can identify infants at risk for the familial forms and spot the tumors early, which helps make their treatment easier and more effective. Because some infants are born with hereditary retinoblastoma tumors that fill the eye or eyes completely, some parents choose prenatal detection and terminate the pregnancy if the retinoblastoma gene is found in the fetus.

In addition, because the gene is missing in people who develop the tumors, scientists have proposed that the gene involved has as its normal function the prevention of the development of the form of cancer. When the gene is lost, therefore, its protective function also is lost. We now use the term "tumor suppressor gene" to describe the situation for several cancer-causing or cancer-predisposing genes. This is a new way of thinking about how cancer might happen; that is, that the absence of a normal gene rather than the necessary presence of an abnormal one can cause cancer. Or, as is the situation for many cases of both retinoblastoma and the *BRCA* genes, there is a predisposing inherited gene mutation that has no effect unless its normal allele also is disrupted by a somatic mutation. This is another extension and refinement of Mendel's principles, based on new evidence. The modification of explanations happens constantly in science as new evidence becomes available. Those insights into the onset of retinoblastoma might lead to new methods of treating cancer in the future.

We don't know the exact function of the missing gene, but it does appear to code for a protein, or perhaps a nonprotein RNA transcript that is involved in binding to DNA. That binding may be an important mechanism to control the events of the cell cycle (cell replication/mitosis). When the process is disrupted, uncontrolled cell division may occur and a cancer develops.

Furthermore, some of the recently discovered suppressor gene products bear a striking resemblance to protein found in certain viral infections, suggesting a link not only between genetics and cancer, but a role for viruses as well. People may have different abilities to react genetically to certain viruses and those differences could lead to the development of cancer at different rates in different groups of people.

IMPRINTING AND RETINOBLASTOMA

The relationship between parental origin and the development of tumors is complex and the re-

search has a long way to go. Some early observations might interest you. For example, a sporadic (nonfamilial) bone cancer known as osteosarcoma most often is associated with and probably due to the preferential loss of the retinoblastoma gene on the maternally derived chromosome 13. Researchers have not found tumors if the mutation is on chromosome 13 from the father. In another cancer that is occasionally familial but usually sporadic (Wilms tumor, a kidney cancer that affects infants or young children), most cases have a loss of all or part of chromosome 11. The cancer develops, however, only when the mutation comes from a deletion on the chromosome 11 derived from the mother. That suggests that the maternal chromosome 11 has some tumor suppressing effect that is not compensated for by the father's chromosome 11. We already have found other examples that will undoubtedly shed more light on the mechanisms of the development of cancer and provide additional extensions of Mendel's principles.

LUNG CANCER: GENETICS OR ENVIRONMENT?

Chris and Andres were on their way to a school assembly. A player from the Seattle Mariners was going to speak on drug and alcohol abuse. "I hope he doesn't go through the cigarette smoking thing again. I've heard it so many times I could give that talk myself," said Scott. Chris knew that Andres was only half serious. They both knew how foolish it

was to smoke cigarettes, but Chris had to admit that it was tiring to hear the same old story and see the same old pictures each year since third grade. But, the player had a surprise for them this year.

"I'm sure you all know by now how dangerous smoking can be for your health," he began. Chris looked at Andres in mild surprise.

"But, what you may not know

is that some of you may have an even greater risk because of your genetic background. Scientists are finding more evidence that certain people react more strongly than others to the chemicals in cigarette smoke. Those people have an even greater chance of developing lung cancer." He continued his presentation and talked about other forms of chemical abuse.

When the assembly was over, however, Andres was curious about the relationship between genetics and cancer. "You know, Chris, when I was a kid," Andres said, "there was a guy named Mr. Sergor in our neighborhood. He would yell at us for taking apples off his trees. But, I knew he didn't really mean it. I always thought that he had climbed trees when he was a kid, and he wished he still could. Anyway, he had worked as a shipbuilder during World War II and had real bad breathing problems. He told us once that a lot of people who worked on those ships had developed problems. I remember that his brother died from lung cancer, and my mom made a big deal of it to all the kids. Anyway, do you think that maybe Mr. Sergor's family had a genetic tendency to develop lung problems?"

Chris had never heard Andres talk about anything that much, so he wasn't surprised when three weeks later Andres gave his biology report, which he titled "Genetic and Environmental Risks for Lung Cancer."

Andres began his report as if he had heard the story of cigarettes and lung cancer twenty times before. But, like the speech at the assembly, it soon got interesting. In addition to the risk from cigarettes, doctors also had found that exposure to asbestos and radon can increase your risk. Asbestos was used widely in insulation and shipbuilding. Mr. Sergor was probably exposed to asbestos.

Radon is a naturally occurring radioactive gas that is found at slightly increased levels in some parts of the country, and it can accumulate in homes that are tightly sealed against air leaks. The risk for contracting lung cancer increases about ten times if you smoke cigarettes, about three times if you work with asbestos for a long time, and about one and a half times if you live for a long time in a house with elevated radon levels.

Even more important, it is now clear that certain individuals produce enzymes in response to the chemicals in cigarette smoke or asbestos. The chemicals become even more toxic to the lung and increase the risk of lung cancer to still higher levels. These enzymes are normal components of our bodies and probably even have an essential function. They can be present as different variants in different people, just as blood groups can vary. For example, Rh negative blood is perfectly normal, but it can create problems in certain situations. The enzyme differences also can create problems in certain situations. Andres concluded by saying that the effects of the different risk factors seem to be independent of one another. To determine your total risk, therefore, you have to multiply the risk for each genetic or environmental factor you have.

Michiku Nakayama raised her hand and asked, "How could you tell if you have one of the enzymes that increase your risk?"

"Well," Andres answered, "right now the test has to be done in a special research lab. But, I got the impression from one article I read that they're working on an easier test that can be done on small amounts of blood." Andres's enthusiasm grew as he continued. "There are actually two different enzymes involved. One of them is called a P450 enzyme, and it breaks down lots of chemicals in the blood—like blood-pressure medications. About sixty percent of the population has a form of the P450 enzyme that increases a risk of lung cancer by five times. The other one is called an AH enzyme, and it also breaks down chemicals in the body—especially hydrocarbon by-products from things like meats cooked in certain ways. About fifty percent of the population has a form of that enzyme that increases their risk of lung cancer risk by ten times."

Chris could tell that Mr. Mosely, the biology teacher, was impressed by Andres's talk. Mr. Mosely seemed glad when they continued the discussion on how genetics and the environment can interact. You can learn more about relative risks and cancer by taking part in the class exercise your teacher will lead.

FUTURE WATCH:
Genetic Engineering

TOOLS OF THE GENETIC ENGINEER

PART A:
IDENTIFYING AND ISOLATING THE hGH GENE

Among the most important and exciting developments in biology during the last thirty years is genetic engineering—the ability to manipulate DNA. Scientists have learned to manipulate genetic material to produce such substances as insulin, human growth hormone, interferon, and hepatitis B vaccine. They use recombinant DNA technology to produce those substances. This activity provides a model for the following steps in that process.

1. Scientists first identify the gene that codes for the production of the protein they want to manufacture (for example, human growth hormone).
2. After they identify the gene, scientists isolate it.
3. Once scientists isolate the gene, they insert it into a plasmid and place it into a host cell, usually a bacterium.
4. The bacteria replicate the gene, multiple copies of the gene become active, and the bacteria begin to produce the protein.

GROWTH HORMONE

Karen had just finished her driver's education course and had gone with her parents to take her road test. After passing her test, she met with a young woman to fill out some forms for her new license. When the woman left her chair to get the forms from a file cabinet, Karen noticed that she was short, about four feet tall, she guessed. After filling out the necessary forms and having her picture taken for her license, Karen left with her parents. On the way home Karen asked her parents if the young woman had the same thing her brother Chris had.

"I don't know. How much do you know about what Chris has?" her father asked.

"I'm not sure," said Karen. "I know he has growth hormone deficiency, because I have always heard you and mom say that when people ask why Chris is so small. And, he goes to the doctor every week to get hormone shots. But, I'm not sure I really understand what the problem is. I know that when Chris was younger he was much shorter than other kids his age. The difference doesn't seem that large, now that he's older."

Chris was now eleven years old, and although they had their disagreements, Karen liked Chris a lot. He was a forward on the Hoover Middle School soccer team, was a rock hound, and always seemed to be in an upbeat mood. Her dad suggested that Karen accompany Chris and him on Chris's next visit to the doctor. "You can speak to Dr. Thompson about what growth hormone is," said Karen's father.

The following Monday, while Chris was being weighed and measured, Karen and her dad sat down with Dr. Thompson. Dr. Thompson was a pediatric endocrinologist, a specialist who cares for children with hormone problems. Many of the children under her care had diabetes, but quite a number also had growth problems. "Well, Karen," Dr. Thompson said, "if you look around at your classmates at school, it is easy to see that people vary in many ways. Height, weight, skin color, eye color, intelligence, and athletic ability are good examples of individual differences. Things that vary in that way are often the result of a combination of many factors. An individual's height, for example, depends on the heights of his or her parents and grandparents, on nutritional factors, and on the function of proteins or hormones from different glands, such as the pituitary, the thyroid, and the adrenals.

"One of the most important of these hormones or proteins is growth hormone, which is made in the pituitary gland," continued Dr. Thompson. Karen remembered from her biology class that the pituitary gland is in the brain. "Abnormalities of growth hormone can occur in several ways," Dr. Thompson explained. "Excess production of growth hormone can result in an individual's being taller than expected and having serious bone and joint problems as he or she gets older. Many of the famous circus giants of the past probably had that condition. Individuals who are shorter than expected also may have growth hormone problems."

"So, Chris has a problem with the pituitary gland," said Karen.

"Yes, his pituitary gland is not making enough growth hormone," said Dr. Thompson. "When Chris was about one and a half years old, your pediatrician, Dr. Kelfer, noticed that Chris wasn't growing as fast as expected. Dr. Kelfer watches the growth of all his patients, from birth, on a growth curve. Chris's curve is shown on this graph, along with the average rate of growth. We did some special tests

on Chris and found he wasn't making growth hormone. Fortunately, we were able to give Chris the growth hormone he needed to have close-to-normal growth."

Karen had a lot more questions about growth hormone, and Dr. Thompson suggested she do some reading about it. Little did Karen know that she would soon become the class expert on genetic engineering because of her interest in the causes of short stature.

The school librarian helped Karen find a book on endocrinology—the study of hormones. She read about growth hormone and found out that it is a protein that contains 191 amino acids. (How many nucleotides of DNA code for the information to construct growth hormone?) This hormone is related to several other hormones involved in regulating growth and metabolism. Some children are deficient in growth hormone and do not grow as fast as expected. Doctors have developed a graph that shows how tall a boy or girl should be at different ages, much like the graph that shows Chris's growth. (Would you use the same curve for both boys and girls? Why?) Some of the children who lacked the normal amount of growth hormone had inherited a deficiency of growth hormone from their parents. That genetic deficiency is a recessive condition. On the average, therefore, one child out of four of such parents is affected. Fortunately, we now can treat these children, like Chris, using growth hormone manufactured through genetic engineering.

That has not always been the case, however. For many years, there was no treatment for people who lacked enough growth hormone. In the 1950s, scientists

found that they could make growth hormone from the pituitaries of people who had died. Unfortunately, not enough people donated their pituitaries, and there was never quite enough growth hormone for all the people who needed it. Only the most seriously affected people were given the growth hormone. Even more serious was the fact that a few of the pituitaries contained a deadly virus that infected some children. Many of them died from the infection.

In the early 1980s, other scientists found a way to get common bacteria to make growth hormone. That was an important breakthrough, because we could produce large quantities of growth hormone, and short children were no longer limited to hormone coming from pituitaries.

Karen was thankful that genetic engineers had come up with a safe form of growth hormone for her brother to use. Because of that discovery her brother was assured of a safe, reliable source of hormone that would allow him to grow at a fairly normal rate.

Materials

For each team of two students:
pop beads: 23 black, 15 white, 28 green, 28 red, 9 yellow
1 piece of string, 15 cm in length

Procedure and Discussion

1. Answer the following questions before you begin building your model of recombinant DNA. In what type of cells would you look for the human growth hormone (hGH) gene? What product, aside from hGH, would tell

you that the gene for hGH is functioning in the tissue you just identified?

2. Now begin the construction of your model. Figure 1 is a flow chart that summarizes the steps. Figure 2 is a portion of messenger RNA (mRNA) for hGH, isolated from pituitary tumor cells. Using pop beads, construct the mRNA strand using the following key:

   black = adenine (A)
   red = cytosine (C)
   white = thymine (T)
   yellow = uracil (U)
   green = guanine (G)

   Scientists have a problem, however, with using this mRNA in genetic engineering. Human DNA—from which mRNA is made—contains base sequences that do not contain information used in the production of protein. This extra information, called introns, was coded into the mRNA you just formed. The mRNA with these introns is called precursor mRNA. We must remove the introns to form mature, functional mRNA.

   Remember that the third step in the recombinant DNA process is the insertion of the isolated gene into a bacterium. Bacterial cells lack the enzymes needed to remove

```
1. Synthesize precursor mRNA (single stranded)
         | Remove intron
         ↓ Reattach exons
2. Produce mature mRNA (single stranded)
         | Assume use of
         | reverse
         ↓ transcriptase
3. Synthesize DNA (single stranded)
         | Add a
         | complementary
         ↓ strand
4. Synthesize double-stranded DNA
      This is the double-stranded gene for hGH
```

FIGURE 1 ▪ Flow chart for the production of the human growth hormone gene.

the introns in the human mRNA you just constructed. Scientists must clip out the introns in this human mRNA before they can engineer a gene into bacteria. (Note: The sequences used in mature, functional mRNA are called exons because they are *ex*pressed into proteins. Introns are *in*tervening sequences.)

3. Your next task is to locate the intron in the human mRNA you constructed. The amino acid sequence for the piece of the hGH gene you are working with is as follows: phenylalanine—glutamic acid—tyrosine—arginine—cysteine—proline—arginine—alanine—cysteine—phenylalanine—glutamic acid.

   Use the genetic code in Table 1 to determine the location of the intron in the precursor mRNA you formed. Determine whether each codon (a sequence of three bases) codes for the specified amino acid. (Remember, those are *mRNA* codons.) For example, the first codon in your mRNA is UUC. That codes for the amino acid phenylalanine, which is the first amino acid in the hGH protein. The second codon is GAA, which codes for glutamic acid—the

KEY																			
black = adenine (A)	1	2	3	4	5	6	7	8	9	10	11	12	13	14	15	16	17	18	19
white = thymine (T)	U	U	C	G	A	A	U	A	C	C	G	A	U	G	U	C	C	G	U
green = guanine (G)	20	21	22	23	24	25	26	27	28	29	30	31	32	33	34	35	36	37	
red = cytosine (C)	A	G	C	A	G	G	G	C	C	U	G	C	U	U	C	G	A	A	
yellow = uracil (U)																			

FIGURE 2 ▪ Portion of messenger RNA for hGH.

second amino acid in the hGH protein. Continue comparing codons and amino acids until you reach bases that do not contain the proper information in the production of hGH. These bases make up the introns. What are the numbers of the bases?

4. "Clip out" the intron by removing the bases in the mRNA. Reconnect the exons. Now you have produced mature, functional mRNA that contains only exons. The next step is to place the hGH gene into a bacterium. What must we do before we take that step?

5. To do that, scientists use an enzyme called reverse transcriptase. That enzyme copies an mRNA molecule into a DNA molecule. It reverses the transcription process, which normally proceeds from DNA to RNA. You will act as the reverse transcriptase and use your mature mRNA to build a coding strand of a DNA molecule. Use the pop beads to build a single DNA strand that is complementary to the mature mRNA strand. (Remember, in RNA, uracil is complementary to adenine.)

6. To make the new DNA double stranded, use pop beads to build the complementary strand of DNA. Use the piece of string to connect the complementary strands as shown in Figure 3. Congratulations! You have constructed an hGH gene. The technique you used to produce a gene from mRNA produces copy DNA or cDNA. Now that you have identified and isolated the hGH gene, what is the next step in the recombinant DNA process?

TABLE 1 The genetic code: codons that specify amino acids.

First Base		Second Base								Third Base
		A or U		G or C		T or A		C or G		
A or U	AAA *UUU* } Phe AAG *UUC* AAT *UUA* } Leu AAC *UUG*		AGA *UCU* AGG *UCC* AGT *UCA* } Ser AGC *UCG*		ATA *UAU* } Tyr ATG *UAC* ATT *UAA* } Stop ATC *UAG*		ACA *UGU* } Cys ACG *UGC* ACT *UGA* } Stop ACC *UGG* } Trp		A or U G or C T or A C or G
G or C	GAA *CUU* GAG *CUC* } Leu GAT *CUA* GAC *CUG*		GGA *CCU* GGG *CCC* GGT *CCA* } Pro GGC *CCG*		GTA *CAU* } His GTG *CAC* GTT *CAA* } Gln GTC *CAG*		GCA *CGU* GCG *CGC* } Arg GCT *CGA* GCC *CGG*		A or U G or C T or A C or G
T or A	TAA *AUU* TAG *AUC* } Ile TAT *AUA* TAC *AUG* } Met-Start		TGA *ACU* TGG *ACC* TGT *ACA* } Thr TGC *ACG*		TTA *AAU* } Asn TTG *AAG* TTT *AAA* } Lys TTC *AAG*		TCA *AGU* } Ser TCG *AGC* TCT *AGA* } Arg TCC *AGG*		A or U G or C T or A C or G
C or G	CAA *GUU* CAG *GUC* } Val CAT *GUA* CAC *GUG*		CGA *GCU* CGG *GCC* CGT *GCA* } Ala CGC *GCG*		CTA *GAU* } Asp CTG *GAC* CTT *GAA* } Glu CTC *GAG*		CCA *GGU* CCG *GGC* CCT *GGA* } Gly CCC *GGG*		A or U G or C T or A C or G

Note: The DNA codons appear in regular type; the complementary RNA codons are in color. A = adenine, C = cytosine, G = guanine, T = thymine, U = uracil (replaces thymine in RNA). In RNA, adenine is complementary to thymine of DNA; uracil is complementary to adenine of DNA; cytosine is complementary to guanine, and vice versa. 'Stop' = chain termination or 'nonsense' codon. 'Start' = signal to begin protein synthesis. The amino acids are abbreviated as follows:

Ala = alanine	Asp = aspartic acid	Glu = glutamic acid	Ile = isoleucine	Met = methionine	Ser = serine	Tyr = tyrosine
Arg = arginine	Cys = cysteine	Gly = glycine	Leu = leucine	Phe = phenylalanine	Thr = threonine	Val = valine
Asn = asparagine	Gln = glutamine	His = histidine	Lys = lysine	Pro = proline	Trp = tryptophan	

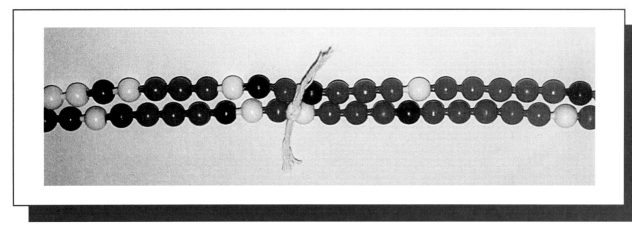

FIGURE 3 ▪ Building double-stranded DNA.

PART B: FORMING RECOMBINANT PLASMIDS

In part A of this activity, you simulated the activities of the genetic engineer in steps 1 and 2 of the recombinant DNA process. Using pop beads as a model, you identified and constructed the hGH gene. In part B, you will investigate how scientists insert the hGH gene into the genetic system of a bacterium and how the bacterium will begin to produce the hGH protein. Figure 4 is a flow chart that summarizes the steps you will follow.

Materials

For each team of two students:
pop beads: 12 black, 12 white, 18 green, 18 red
3 pieces of string or masking tape, 15 cm in length

Procedure and Discussion

Some bacterial DNA exists in small circular pieces called plasmids. When genetic engineers use plasmids to introduce new genes into bacteria, such a tool is called a vector. We can compare the genetic products involved in the recombinant DNA process to a manufacturing plant:

Manufacturing Plant	Recombinant DNA Process
raw materials	cDNA or hGH gene
machinery	plasmid
factory	bacteria
product	hGH protein

1. The first problem is to get the hGH gene (the cDNA you synthesized in part A) into a bacterial plasmid. To solve that problem, you must build a plasmid. Use your pop beads and the DNA sequence in Figure 5 to construct a double-stranded DNA sequence. Use the color key from part A for the bases.

This is a model of a plasmid. Plasmids are circular. To make your model more accurate, connect the two ends of each strand as shown in Figure 6. Mark the connections with a small piece of string or tape. Make strand A the outside circle. You now have the raw material (the hGH gene) and the machinery (the plasmid) of our recombinant DNA process. What is the next step in the process?

1. Construct a bacterial plasmid (form the strands into circles, with strand B on the inside)

2. Identify a restriction enzyme that will cut the cDNA and the plasmid DNA

3. Cut the cDNA with the enzyme

4. Cut the plasmid with the enzyme

5. Form a recombinant plasmid

FIGURE 4 ▪ Flow chart for the formation of a recombinant plasmid.

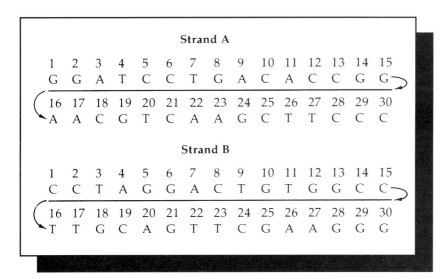

Strand A

1	2	3	4	5	6	7	8	9	10	11	12	13	14	15
G	G	A	T	C	C	T	G	A	C	A	C	C	G	G

16	17	18	19	20	21	22	23	24	25	26	27	28	29	30
A	A	C	G	T	C	A	A	G	C	T	T	C	C	C

Strand B

1	2	3	4	5	6	7	8	9	10	11	12	13	14	15
C	C	T	A	G	G	A	C	T	G	T	G	G	C	C

16	17	18	19	20	21	22	23	24	25	26	27	28	29	30
T	T	G	C	A	G	T	T	C	G	A	A	G	G	G

FIGURE 5 ▪ DNA sequences for the bacterial plasmid.

2. Genetic engineers use enzymatic scissors—called restriction enzymes—to cut open DNA sequences at specific locations. Each enzyme recognizes only a certain sequence of bases, and cuts within that sequence. When the DNA is cut, the ends become "sticky." They can attach to other sticky ends that have been cut with the same enzymes and, therefore, have complementary nucleotide sequences. Figure 7 shows three restriction enzymes, their DNA recognition sequences, DNA cutting sites, and the sticky ends they create.

Imagine that the model of the hGH gene you made in part A is only a small part of a long DNA molecule. Your problem now is to cut out this hGH gene from this long DNA molecule and insert it into the plasmid that you produced in step 1. To do that, you must use a restriction enzyme that will yield sticky ends in the hGH gene that can bind with complementary sticky ends in the plasmid.

Examine your model of the hGH gene, which you constructed in part A of this activity. Which of the restriction enzymes in Figure 7 can you use to remove the hGH gene from the complete DNA molecule? Why did you select

that enzyme? Now examine your model of the plasmid and locate a similar DNA sequence cutting site. Which restriction enzyme can you use to cut open the plasmid? Why?

3. Now act as the restriction enzyme you have chosen and cut the hGH gene at the proper cutting site. What sticky ends have you made in the hGH gene?

4. Acting as the same restriction enzyme, cut open the plasmid. What sticky ends have you made in the plasmid? Compare the sticky ends of the hGH gene and the plasmid. What do you observe?

5. In nature and in the laboratory, complementary sticky ends of a gene and a plasmid are "pasted" back together by enzymes called ligases. Perform the function of ligase by

FIGURE 6 ▪ Model of a plasmid.

Enzyme Name	DNA Recognition Sequence	DNA Cutting Site	"Sticky Ends"
*Bam*H1	GGATCC CCTAGG	—G | GATC C— —C CTAG | G—	—G GATCC— —CCTAG G—
*Hin*dIII	AAGCTT TTCGAA	—A | AGCT T— —T TCGA | A—	—A AGCTT— —TTCGA A—
*Hpa*II	CCGG GGCC	—C | CG G— —G GC | C—	—C CGG— —GGC C—

FIGURE 7 ▪ Examples of restriction enzymes.

connecting the sticky ends of your gene model and your model plasmid together. Use small pieces of tape or string to identify the connections (called splices).

Check your own work. What should you look for to determine whether your gene insertion into the plasmid is correct? Assuming your work has been accurate up to this point, you now have produced a recombinant plasmid. You have recombined the cDNA from the hGH gene and the DNA from the plasmid.

6. What is the next step in the recombinant DNA process?
7. Review the tools of the genetic engineer you used to engineer bacteria whose product is hGH.
8. What would have resulted if you had cut the hGH gene near the middle rather than at the ends? 🐾

WHAT ARE RFLPS?

Linda Fulcher and her husband Glenn were a bit nervous as they met with Beth Greendale, the genetic counselor at Children's Hospital. The Fulchers, a young Black couple from Miami, had been married for three years, and now they wanted to start a family.

Like most young people, the Fulchers were healthy. They assumed that any children they might have would be healthy, too. But, when they heard at their church about the screening program for sickle-cell trait, they thought they should look into it.

They found that about 1 in 625 Black children in the United States is born with sickle-cell disease each year. When Beth Greendale gave her presentation to the church group as part of the screening program, she said that sickle-cell disease is caused by a problem with hemoglobin, the protein that carries oxygen in the blood. As a result of the abnormal hemoglobin, the blood cells take on a sickled shape, as shown in Figure 1. The cells are long and rigid, instead of round and flexible. People who have the disorder experience severe pain in their long bones and joints and are susceptible to infections. Some do not live beyond childhood and some even die as infants of overwhelming bacterial infection. With good medical care, however, affected people can live reasonably normal lives into adulthood.

Sickle-cell disease, Beth said, is a recessive trait. That means affected individuals have two abnormal genes, one inherited from each parent. Some people, Beth said about one in ten people in the Black population, have one sickle-cell gene. They are carriers and have sickle-cell trait. They are not anemic and have no health problems related to the sickle gene, but if they produce children with another carrier, there is a twenty-five

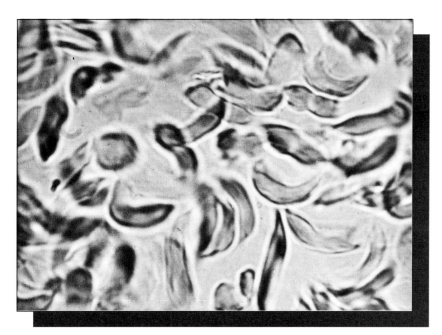

FIGURE 1 ▪ Sickled cells.

percent chance that each child will have sickle-cell disease.

Beth had explained that each major ethnic group seems to have at least one genetic disorder that occurs more frequently in that population. Blacks, who trace their ancestry to Africa, and people who trace their ancestry to certain regions around the Mediterranean Sea, are at increased risk for sickle-cell disease.

Linda and Glenn decided to have themselves tested for sickle-cell trait. When the results showed they both had sickle-cell trait, they called Beth at the genetic counseling clinic at Children's Hospital to discuss family planning. Beth told them that there is a very accurate prenatal test that can determine whether the developing fetus has sickle-cell disease, sickle-cell trait, or normal hemoglobin. Linda and Glenn listened as Beth introduced them to RFLPs.

Red blood cells are filled with a protein called hemoglobin. Hemoglobin carries oxygen to cells that are actively working. In the late 1940s, Dr. Linus Pauling suggested that sickle-cell disease might be caused by an abnormality in hemoglobin.

Many years of research revealed that the DNA that codes for the hemoglobin molecule has a mutation in individuals who have sickle hemoglobin. The mutation changes one amino acid in normal hemoglobin (glutamic acid) to a different amino acid (valine) in sickle hemoglobin. That change gives the red blood cells a different shape, which causes the problems seen in affected individuals.

Materials

For each team of two students: pop beads: 6 black, 30 white, 12 green, 22 red

Procedure and Discussion

1. Strands A and B in Figure 2 represent two single strands of DNA. Using pop beads, one team member will make a model of strand A and the other member will make a model of strand B. Compare the two strands and note any differences.

    Key: black = adenine (A)
    white = thymine (T)
    green = guanine (G)
    red = cytosine (C)

2. Use the genetic code in Table 1 to determine the amino acids that are coded for by bases 17, 18, and 19; by bases 20, 21, and 22; and by bases 23, 24, and 25 in both strands. Note any differences.

3. Normal hemoglobin has the amino acid glutamic acid in its DNA strand and sickle hemoglobin has the amino acid valine in its DNA strand. Based on that information, what can you predict about an individual whose genetic makeup contains strand A? strand B?

4. Restriction enzymes recognize specific DNA sequences and cut the DNA at a specific place in each sequence. The restriction enzyme *Mst*II recognizes the DNA sequence GGTCTCC and cuts the DNA

    ↑ cut

    between the first T and the first C of that sequence, reading from left to right. Identify the restriction sites in each strand and separate the pop beads at the appropriate place. What is the length of the fragments (how many bases) when you cut strand A with *Mst*II? What is the length of the fragments (how many

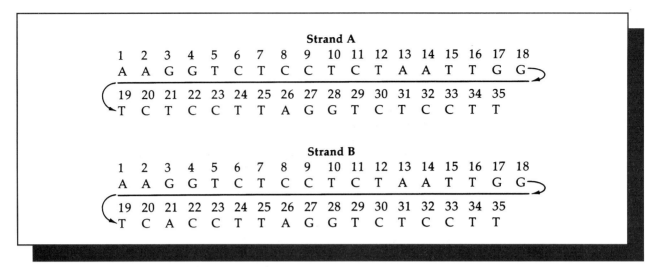

FIGURE 2 ■ Two single strands of DNA.

TABLE 1 The genetic code.										
					Second Base					
First Base	A or U		G or C		T or A		C or G			Third Base
A or U	AAA UUU | AAG UUC \} Phe	AAT UUA | AAC UUG \} Leu	AGA UCU | AGG UCC | AGT UCA | AGC UCG \} Ser		ATA UAU | ATG UAC \} Tyr	ATT UAA | ATC UAG \} Stop	ACA UGU | ACG UGC \} Cys	ACT UGA \} Stop | ACC UGG \} Trp		A or U | G or C | T or A | C or G
G or C	GAA CUU | GAG CUC | GAT CUA | GAC CUG \} Leu		GGA CCU | GGG CCC | GGT CCA | GGC CCG \} Pro		GTA CAU | GTG CAC \} His	GTT CAA | GTC CAG \} Gln	GCA CGU | GCG CGC | GCT CGA | GCC CGG \} Arg			A or U | G or C | T or A | C or G
T or A	TAA AUU | TAG AUC \} Ile | TAT AUA | TAC AUG \} Met-Start		TGA ACU | TGG ACC | TGT ACA | TGC ACG \} Thr		TTA AAU | TTG AAG \} Asn | TTT AAA | TTC AAG \} Lys		TCA AGU | TCG AGC \} Ser | TCT AGA | TCC AGG \} Arg			A or U | G or C | T or A | C or G
C or G	CAA GUU | CAG GUC | CAT GUA | CAC GUG \} Val		CGA GCU | CGG GCC | CGT GCA | CGC GCG \} Ala		CTA GAU | CTG GAC \} Asp | CTT GAA | CTC GAG \} Glu		CCA GGU | CCG GGC | CCT GGA | CCC GGG \} Gly			A or U | G or C | T or A | C or G

Note: The DNA codons appear in regular type; the complementary RNA codons are in color. A = adenine, C = cytosine, G = guanine, T = thymine, U = uracil (replaces thymine in RNA). In RNA, adenine is complementary to thymine of DNA; uracil is complementary to adenine of DNA; cytosine is complementary to guanine, and vice versa. 'Stop' = chain termination or 'nonsense' codon. 'Start' = signal to begin protein synthesis. The amino acids are abbreviated as follows:

Ala = alanine   Asp = aspartic acid   Glu = glutamic acid   Ile = isoleucine   Met = methionine   Ser = serine   Tyr = tyrosine
Arg = arginine   Cys = cysteine   Gly = glycine   Leu = leucine   Phe = phenylalanine Thr = threonine Val = valine
Asn = asparagine Gln = glutamine   His = histidine   Lys = lysine   Pro = proline   Trp = tryptophan

bases) when you cut strand B with *Mst*II? How could you use *Mst*II to distinguish sickle hemoglobin from normal hemoglobin?

5. Assume that Ms. Fulcher is pregnant and undergoes prenatal diagnosis. The genetic counselor shows Ms. Fulcher and her husband the results of the DNA test. The following are three possible results of the prenatal diagnosis. In each case, give the diagnosis for the developing fetus: fragment sizes = 5, 14, 10, 6; fragment sizes = 5, 14, 10, 24, 6; fragment sizes = 5, 24, 6.

6. The title of this activity asks: What are RFLPs? You have just used RFLPs to test for the presence of sickle hemoglobin. RFLP stands for restriction fragment length polymorphism. Use your knowledge of restriction enzymes and variations in DNA sequences to write a paragraph that explains the name restriction fragment length polymorphisms. (Not-so-helpful hint: polymorphism means many forms, in this case, many different lengths.)

7. If you used the process from step 4 on the DNA of an individual who is heterozygous for an autosomal dominant disorder such as Huntington disease or on the DNA of an X-linked disorder in a male, what patterns of RFLPs might you look for in the two cases? How would heterogeneity—different mutations causing the same disorder—affect a diagnosis made on the basis of RFLPs? What tissues can we use for that type of diagnosis?

The Human Genome Project

Mapping and Sequencing Human Genes

Suppose that it becomes possible for geneticists to identify the genes that predispose certain people to heart disease, or those genes that will cause individuals to become blind as they age. If you knew what genes you have and what genetic disorders they might cause, would it affect your choice of careers? Would it affect your decision to marry or begin a family? Suppose your potential employer also had access to your genetic information. Could the fact that an employer had knowledge about your genetic information result in job discrimination? If you knew that some genes in your unborn baby might cause him or her to suffer from impaired mental or physical performance, would that affect your decision to carry the fetus to term? Those are some of the issues raised by the Human Genome Project, or HGP. Decisions such as those likely will confront society as the HGP generates information on the content of the human genome.

Human cells that have nuclei (except for eggs and sperm) normally contain twenty-two pairs of autosomes, plus two sex chromo-somes (XX or XY). Those cells also contain mitochondria, the powerhouses of the cell. Each mitochondrion has a small circular chromosome. The total genome for any individual includes all of those components.

The primary goal the HGP is to uncover the genetic information contained in the human chromosomes. For the purposes of the HGP, genome means one each of the different chromosomes—twenty-two autosomes (not the diploid number of forty-four), plus X, plus Y, plus a mitochondrial chromosome. This genetic material includes approximately three billion base pairs of DNA, containing fifty thousand to eighty thousand genes (see box). Researchers will map each chromosome and then attempt to determine the DNA sequence of each gene. The HGP will not provide the DNA sequence from one individual, but rather a collection of sequences from different people. The knowledge and technology that result from the HGP eventually may allow researchers to describe the total genome of *Homo sapiens*.

The project, which began in 1990 and is funded in the United States primarily by the Department of Energy and the National Institutes of Health, will take about fifteen years and will cost approximately three billion dollars to complete. About six other countries also are involved in the HGP. In addition to groups that address the science and data-storage issues, there are groups that address the social, ethical, and legal implications of genome research.

What Might We Learn?

Biologists already have some ideas about the possible results of the HGP, although there may be surprises as well. Geneticists are attempting to isolate genes responsible for certain genetic disorders. Recent successes include the identification of the genes responsible for Duchenne muscular dystrophy and cystic fibrosis—two serious single-gene disorders—and certain forms of breast cancer. By isolating such genes, biologists can learn about the structure of the gene's corresponding protein

How Much Is Three Billion?

To get some idea about how much three billion is, we can compare the human genome to the planet Earth, as shown in Figure 1. There are about five billion people on the planet, and there are about three billion base pairs of DNA in the human genome. In this case, each of our twenty-three pairs of chromosomes would then be similar to a large country—the United States, for example. A chromosome band would then be similar to a state—like Ohio. Each gene would then be the same as a city or town within that state, such as Columbus. An exon within that gene might then be like a neighborhood, and the individual base pairs would then correspond to individual people in that town.

If we wanted to identify a particular person on the planet, we would need very detailed information about the person's address—a kind of "map position." Similarly, if we want to find the causes for some kinds of genetic disorders, we must identify their exact position on the genetic map. The genome project is designed to establish that kind of very detailed map. In addition, the project also will tell us about human variability, lead to the development of new biomedical technology, and open doors to new ways of diagnosing and treating inherited disorders.

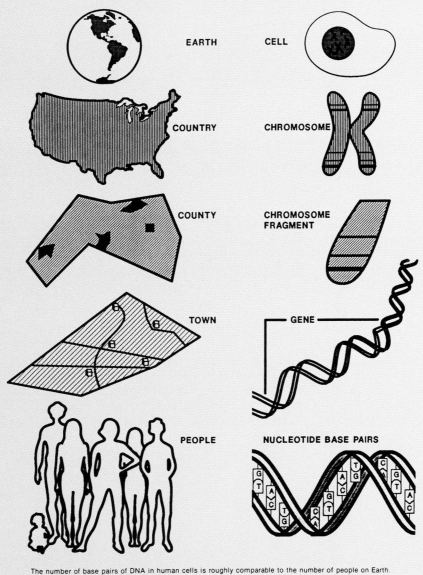

EARTH   CELL

COUNTRY   CHROMOSOME

COUNTY   CHROMOSOME FRAGMENT

TOWN   GENE

PEOPLE   NUCLEOTIDE BASE PAIRS

The number of base pairs of DNA in human cells is roughly comparable to the number of people on Earth. The scale of genetic mapping efforts can be compared to population maps, with chromosomes (50 to 250 million base pairs) analogous to nations, and genes (thousands to millions of base pairs) to towns.

FIGURE 1 ■ Comparative scale of gene mapping.

and the cause of the disorder. That knowledge could lead to better medical management of such diseases.

Genome information also can indicate whether an individual is likely to develop a disease in the future. For example, if the gene for Huntington disease is present, it is a near certainty that symptoms eventually will occur. The HGP also may help predict which individuals have an increased susceptibility to disorders such as heart disease, cancer, or diabetes. Because those disorders result from complex interactions between genes and the environment, there is no certainty that symptoms will develop. The risk, however, is greater for certain individuals than for the general population. Biologists also are hopeful that the HGP will shed light on basic biological questions such as human evolution, development, and gene regulation.

In addition, the HGP also is responsible for the development of new technologies that will have an impact outside the HGP itself. One such technology is sophisticated data-management systems and computers to handle the enormous amount of information generated by the HGP.

WHAT ARE THE CRITICISMS?

As with many large projects, critics have expressed several concerns about the HGP. One major criticism is that the high cost is not justified because we will not learn very much about what it is to be human simply by sequencing all the genes in the human genome. We also must take into account the numerous influences of the environment. Other critics maintain that funding such a large project takes scarce resources away from individual researchers who may study more efficiently certain areas of particular scientific or medical interest. In addition, allocating funds for research on genetic disorders takes money away from programs that help relieve other causes of ill health. Perhaps the genome money would be better spent, for example, on prenatal care for all pregnant women, on nutrition and basic health care for children living in poverty, or on AIDS research.

Some critics suggest that the ability to diagnose a genetic disorder before any treatment is available does more harm than good because it creates anxiety and frustration. Even in the absence of new treatments, however, the HGP may make diagnosis possible before the onset of symptoms and, thus, make the management of the disorder more effective. In addition, improved knowledge of one's genetic background can provide couples with a broader range of options for family planning.

Other critics object to the HGP because they believe we do not have the ability to regulate the ultimate outcomes. Some critics do not feel that just because scientists *can* do this science, they *ought* to. Those critics point to the development of atomic weapons and argue that the science that led to their development caused far more problems than it resolved.

It is possible that the HGP will allow the determination of personal genetic profiles—an individual's genome data as opposed to genome data that are reflective of the general human population. There are many unanswered questions about the use of such individual data. Who will have access to such personal genome data? Should employers or insurance companies be permitted to use the data to discriminate against individuals? How can we ensure the privacy of the very large databases to which many individuals and organizations have access? Should information such as disease-carrier status or paternity information be provided to family members who did not seek the information themselves?

Some critics feel that the genome project gives too much emphasis to the genetic component of human characteristics while ignoring the environmental components. They worry that that emphasis will cause society to blame people for having "bad" genes while failing to improve environmental and societal factors that might help such individuals escape the effects of those genes. What do you think about the Human Genome Project?

GENETIC TECHNOLOGY IN BASIC RESEARCH: EVOLUTION THEORY

Biological classification is the process by which biologists assign organisms to specific groups. The group may be as large and encompassing as a kingdom, such as animalia, or as small and limiting as a species, such as *Homo sapiens*, the species to which we assign humans.

Classification is not merely a method of assigning organisms to groups so that biologists can keep track of them. The groups to which organisms are assigned also imply the degree of relatedness between those organisms. Each classification is an active hypothesis about the evolutionary history of the species in question and about its relationship to other species. Organisms that share smaller groups of classification are more closely related than are organisms that share only large groups.

Until the mid 1970s, classification was based largely on the comparison of observable structures in a given organism. For example, taxonomists might compare the structure of forelimbs in mammals. In recent years, they also have been able to compare the structure of certain proteins.

Modern research techniques allow biologists to compare the DNA that codes for certain proteins and to make predictions about the relatedness of the organisms from which the DNA was taken. This investigation shows you a model of that technology. It might be helpful for you to review the relationship between DNA and proteins in the appendix "The Genetic Code and Production of Protein."

PART A: COMPARING DNA STRANDS

Materials

For each team of four students:
paper clips: 35 each of black, white, green, red
4 small pieces of paper
4 short pieces of masking tape

Procedure and Discussion

1. Which organisms in Table 1 share the most groupings? On the basis of the data in the table, which organisms are least closely related?

TABLE 1 Classification of four species in the kingdom animalia.				
Level of classification	Human	Gorilla	Southern Leopard Frog	Katydid
Phylum	Chordata	Chordata	Chordata	Arthropoda
Subphylum	Vertebrata	Vertebrata	Vertebrata	
Class	Mammalia	Mammalia	Amphibia	Insecta
Subclass	Eutheria	Eutheria		
Order	Primates	Primates	Salientia	Orthoptera
Suborder	Anthropoidea	Anthropoidea		
Family	Hominidae	Pongidae	Ranidae	Tettigoniidae
Subfamily			Ranidae	
Genus	Homo	Gorilla	Rana	Scudderia
Species	Homo sapiens	Gorilla gorilla	Rana pipiens	Scudderia furcata
Subspecies			Rana pipiens sphenocephala	Scudderia furcata furcata

2. Working in teams of four, "synthesize" the DNA strands as follows. The paper clips will represent the four bases of DNA according to the following key:

black = adenine (A)   green = guanine (G)
white = thymine (T)   red = cytosine (C)

Synthesize DNA strands by hooking together paper clips of the appropriate colors. Stretch the string of paper clips out on the lab table with position 1 on the left. Label a small piece of paper as indicated and tape it to the position 1 paper clip. Each strand will represent a small section (twenty bases) of a gene that codes for the protein hemoglobin.

3. **Team member 1**: Synthesize the following piece of DNA by hooking together paper clips in this sequence.

position 1   position 20
↓   ↓
A–G–G–C–A–T–A–A–A–C–C–A–A–C–C–G–A–T–T–A

Label the strand "human DNA." This will represent a small section (20 bases) of the gene that codes for the protein hemoglobin in humans.

4. **Team member 2**: Synthesize the following strand:

position 1   position 20
↓   ↓
T–C–C–G–G–G–G–A–A–G–G–T–T–G–G–C–T–A–A–T

Label this strand "chimpanzee cDNA." (cDNA stands for complementary DNA.) cDNA is a single strand of DNA that will match up with its partner strand. Remember, the bases in DNA are complementary. That is, adenine (A) will always pair with thymine (T), and cytosine (C) will always pair with guanine (G). This cDNA was made from the gene that codes for chimpanzee hemoglobin.

5. **Team member 3**: Synthesize the following piece of cDNA:

position 1   position 20
↓   ↓
T–C–C–G–G–G–G–A–A–G–G–T–T–G–G–T–C–C–G–G

Label this strand "gorilla cDNA." This cDNA strand was made from the gene that codes for gorilla hemoglobin.

6. **Team member 4**: Synthesize the following piece of DNA:

position 1   position 20
↓   ↓
A–G–G–C–C–G–G–C–T–C–C–A–A–C–C–A–G–G–C–C

Label this strand "hypothetical common ancestor DNA." You will use this strand in part B of the activity.

7. Compare the sequences of the human DNA and the chimpanzee cDNA. Match the human DNA and the chimpanzee cDNA base by base (paper clip by paper clip). Remember, black (adenine) must always pair with white (thymine), and green (guanine) must always pair with red (cytosine). If the bases are complementary (that is, if the colors match correctly), allow the clips to remain touching, as shown in Figure 1. If the bases are not

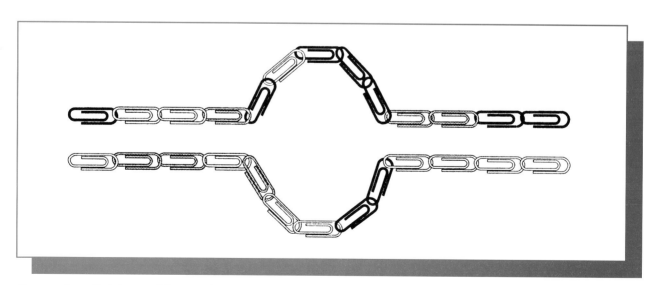

FIGURE 1 ▪ Pairing and loops of complementary DNA strands.

TABLE 2 Hybridization data.		
Human DNA hybridized to:	Chimpanzee cDNA	Gorilla cDNA
Number of loops		
Number of differences		

complementary, separate the clips slightly to form a loop, as shown in the figure. Count the number of loops. Also count the total number of bases that do not match. Record the data in the appropriate columns in Table 2.

8. Repeat step 7, using the gorilla cDNA and the human DNA. Enter the data in the appropriate columns in Table 2.

9. Based on the data you collected for this one protein, is the gorilla gene or the chimpanzee gene more similar to the human gene?

10. Again, based on the data you collected for this one protein, does the gorilla or the chimpanzee seem more closely related to humans?

11. Do the data prove that your answer to the second question in step 1 is correct? Why or why not?

12. How might you provide additional support for the hypothesis you proposed in your answer to question 10?

PART B: AN EVOLUTIONARY PUZZLE

Biologists have determined that mutations in DNA occur at a regular rate. They can use that rate as a "molecular clock" to predict how long ago in evolutionary history two organisms began to separate from a common ancestor. In this part of the activity, you will use your paper-clip model to provide data to support one of two hypotheses about a common ancestor for humans, chimpanzees, and gorillas.

Procedure and Discussion

1. Read the information in Figure 2 about a current debate among scientists who study human evolution.

2. Use your DNA model and the DNA sequences from part A to investigate this debate. First, use your human DNA as a guide to synthesize a human cDNA. Do that as a team. Make certain that the human cDNA is complementary to the human DNA strand.

3. Assume that the DNA for the hypothetical common ancestor you synthesized in part A is DNA for hemoglobin extracted from a hypothetical common ancestor. Now, match all three samples of cDNA (gorilla, human, and chimpanzee) with the common ancestor DNA, one sample at a time. Allow the paper clips to touch where the bases match correctly; form loops where the bases do not match. Record your data in Table 3.

4. Which cDNA is most similar to the common ancestor DNA?

5. Which cDNAs are most similar to each other in their patterns of matching and looping when matched to the common ancestor DNA?

6. Which model in the evolutionary debate described in Figure 2 do your data support?

7. Do your findings prove that that model is the correct one? Why or why not?

TABLE 3 Hybridization data: common ancestor.			
Common ancestor DNA hybridized to:	Human cDNA	Chimpanzee cDNA	Gorilla cDNA
Number of loops			
Number of differences			

Most scientists agree that humans, gorillas, and chimpanzees shared a common ancestor at one time in evolutionary history. One group, however, thinks the fossil record shows that gorillas, chimpanzees, and humans split from one common ancestor at the same time. Here is their model for this split:

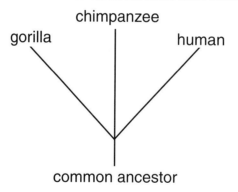

A second group thinks the fossil record shows that there were two splits. In the first, gorillas split from the common ancestor. Humans and chimpanzees then shared another common ancestor for perhaps two million years. They then split again and evolved into their present states. Here is their model for that pattern of splitting:

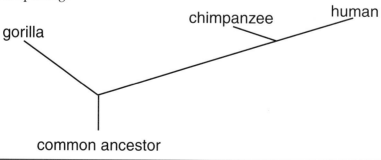

FIGURE 2 ▪ Two models that illustrate common ancestry.

THE MYSTERY OF THE MISSING "FAITH" DIAMOND

Lord Myron Thistleroot was escorting his guests from the United States on a sight-seeing trip of his palatial Chelsea home. They had already spent two hours viewing the many rooms and countless works of art that adorned the walls of these rooms. Lord Thistleroot was very excited; he was saving the best for last. Just two days earlier he had purchased the second-most-perfect diamond in the United Kingdom, the famous "Faith" diamond. His security staff had placed the diamond in a temporary glass display case just that morning. There was nothing else in the room.

At last—the final room for his friends to admire. With a flourish, Thistleroot threw open the door and with a wave of his arm directed their attention to the center of the room. He examined their faces for looks of awe and amazement. Instead, he saw only bewilderment and uneasiness. He looked into the room and shrieked, "My diamond is gone!"

A hasty call to the local constabulary brought detective Sir Lock Romes immediately to the scene of the crime. His careful examination of the room turned up only one clue: the thief had cut himself or herself on the broken glass of the display case, and there were small drops of blood on the broken glass. Sir Lock immediately asked Lord Thistleroot to assemble all his family, friends, and servants in the ballroom. Fifteen minutes later, there were forty-two people gathered for Sir Lock's inspection. He planned to examine the hands of each of the forty-two people and look for reasonably fresh cuts. To his amazement, three out of the forty-two people, Joe the cook, Fred the butler, and Leah the carpenter, had cuts on their hands. Any of the cuts could have been caused by breaking the glass on the display case. He questioned the three, and each seemed to have a perfectly good explanation for his or her injury.

Sir Lock knew he was at a dead end. He needed some help from a relatively new department in the constabulary—the DNA fingerprinting office. Sir Lock's deputy called Dr. Judy Alexander, the director of that office. When she arrived, she quickly took blood samples from the three suspects, and a sample of blood from the broken display case.

Let's leave our story for a moment and look at DNA fingerprinting. In the early 1970s, biologists discovered that certain bacterial enzymes, called restriction enzymes, can recognize specific nucleotide sequences in double-stranded DNA. The enzymes can cut both strands at this site, called a restriction site. An example is the enzyme *Hind*III, which recognizes and cuts the

cut
↓

nucleotide sequence A A G C T T. Any time that sequence occurs in DNA, it will be
T T C G A A
↓
cut

cut by *Hind*III.

If all the DNA from a human cell is chopped up with a restriction enzyme, the result will be thousands of DNA fragments. A special process called gel electro-phoresis separates these fragments by size. The bands that result look a bit like price codes on supermarket items. The position of each band on the gel indicates the length of the sequence fragment, as shown in Figure 1.

The pattern of bands—the DNA fingerprint—varies from person to person when DNA from different people is cut with the same restriction enzyme. Why? Except for identical twins, the likelihood that the bands from two individuals will match perfectly may be as low as 1 in 25 billion.* That makes DNA fingerprinting a highly accurate method of identification.

Now, back to the mystery of the missing "Faith" diamond. Dr. Alexander took the four blood samples to her laboratory and processed them to extract samples of the DNA from the white blood cells. You will now simulate Dr. Alexander's work.

Materials

For each team of two students:
pop beads: 28 black, 28 white, 32 red, 32 green
1 index card (restriction enzyme)

Procedure

1. Colored pop beads will represent the four bases of DNA according to the following key:
   black = adenine (A)
   green = guanine (G)
   white = thymine (T)
   red = cytosine (C)

---

*Helminen, P., et al. 1988. "Application of DNA Fingerprints to Paternity Determinations." *Lancet I* pp. 674–76.

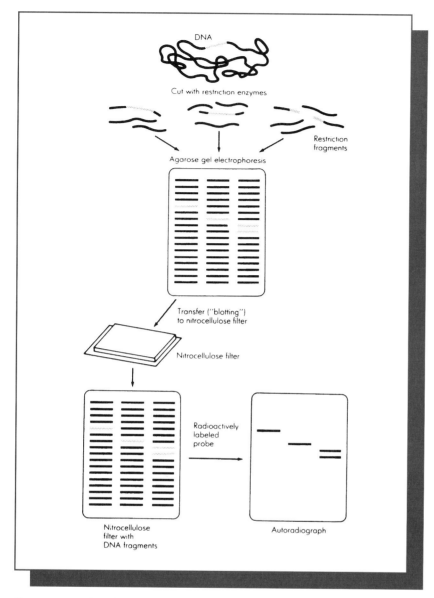

FIGURE 1 ■ Southern blotting. This technique for transferring DNA fragments from gel electrophoresis (in which the fragments are separated according to their size) to a nitrocellulose filter (to which they become bound, still in the same pattern) makes it possible for scientists to pick out specific fragments of DNA. Once the fragments (represented here by lines, but more often in the form of a continuous smear of DNA) are on the nitrocellulose filter, they can be exposed to radioactive probes that will hybridize (stick) to any complementary sequence. The hybridized fragments will then give off radioactive signals that can be made visible on X-ray film in a procedure known as autoradiography.

Your teacher will give each team the base sequences for a DNA strand and its complementary strand. Synthesize both strands by using the pop beads.

2. Next, your teacher will give each team an index card that represents the restriction enzyme *Bam*HI. The enzyme name, the DNA recognition site, and the cutting site are identified on the card.

3. The A-T base pair should be to your left. Place the restriction-enzyme card underneath the DNA at the extreme left of strands and move the card slowly to the right. If you find a DNA sequence in the strands that is identical to the sequence on the index card, stop and "cut" the DNA sequence by taking apart the pop beads or both strands at the appropriate places.

4. If, as a result of step 3, you have cut the DNA strands into fragments, count the number of bases in the fragments in *each* strand. Always count from left to right. Record the number of bases in Table 1. Get data from the teams that examined the DNA from the other two suspects and record those data in Table 1.

5. Repeat the procedure with the second set of DNA sequences given to you by your teacher. This is DNA from the blood on the broken case. Record those data in Table 1.

INTERPRETING THE RESULTS

1. Examine the data you recorded in Table 1 and determine which of the three suspects—Fred, Joe, or Leah—seems to be guilty on the basis of that evidence.

2. What part of the actual process of DNA fingerprinting does step 4 represent?

3. What causes the variation in DNA fragment sizes between individuals?

TABLE 1 DNA fragment lengths from four blood samples.			
	Number of bases in fragment number 1	Number of bases in fragment number 2	Number of bases in fragment number 3 (if present)
DNA from the broken case	strand A = strand B =	strand A = strand B =	strand A = strand B =
DNA from Joe the cook	strand A = strand B =	strand A = strand B =	strand A = strand B =
DNA from Fred the butler	strand A = strand B =	strand A = strand B =	strand A = strand B =
DNA from Leah the carpenter	strand A = strand B =	strand A = strand B =	strand A = strand B =

THE CASE OF NATHANIEL WU

You probably have had some experience, either directly or indirectly, with discrimination and looking for a job. In this activity, you will analyze a dilemma that involves a young man's career and his attempt to secure a challenging job even though his genetic profile may indicate a potential health problem of which he had been unaware. The dilemma involves ethics—the study of what is right and wrong and what is good or bad for individuals, institutions, and society. Although ethics as a field of study may sound obscure, you make ethical decisions all the time—for example, do you ever park in a parking space reserved for handicapped people? Ethics establishes what we should or should not do on the basis of well-reasoned arguments. Well-

reasoned arguments are clearly stated reasons that together justify a conclusion. Well-reasoned ethical arguments can help us see the difference between justified and unjustified discrimination.

PART A: GATHERING AND EVALUATING INFORMATION

Why might it be justified to exclude a person who has a high predisposition to heart disease from a career as a pilot? What about excluding a person who has blue eyes, or who is Hispanic or Jewish? Discrimination means identifying differences, something we do all the time. When might employment discrimination be justified? When might employment discrimination be

unjustified? *The Case of Nathaniel Wu* involves job discrimination based on the type of information generated by research in human genetics.

The Case of Nathaniel Wu

Nathaniel Wu is a top-notch microbiologist. Now thirty years old, he has spent several years working in one of the best research laboratories in the world and has developed an excellent reputation as a creative researcher and hard worker. Following the birth of their son six months ago, Nathaniel and his twenty-nine-year-old wife decided it was time for Nathaniel to seek a job that could help them settle down and become financially secure. Thus, it was with great interest that Nathaniel read the following advertisement in a scientific journal:

Intercontinental Pharmaceutical Corporation (IPC) of New Jersey is seeking highly qualified scientists to join a unique research team. IPC is prepared to invest up to $100 million to establish and support a team of researchers to conduct creative research to find new treatments and cures for diseases ranging from AIDS to heart disease and the common cold. Because IPC will commit $100 million to this research effort, it will require those selected for this special project to commit to a long-term employment contract. Interested applicants should send information to Dr. Anna Peters at IPC.

That was the type of job Nathaniel had always hoped for, and he applied immediately. Before long, IPC invited Nathaniel to come to its headquarters and interview for a position on its special research team.

Dr. Peters, the head of the research committee, led a series of interviews with Nathaniel and three other qualified applicants. Although the other three applicants also were well qualified, they did not seem to share Nathaniel's determination and drive. She listened carefully when Nathaniel presented his latest research findings to IPC scientists. They, too, were impressed with Nathaniel's knowledge, research skills, and potential for contributing to the special goals of the research team. Nathaniel sounded like the type of applicant who could have a long and productive career with IPC, and he seemed to be the sort of team player IPC was seeking.

Because Nathaniel was a top-quality research scientist, there was a high likelihood that his knowledge and research efforts might result in the type of discoveries for new drugs and treatments that were the goal of this special research project. Such discoveries and products could improve the quality of life for countless individuals and dramatically increase earnings for IPC. The investment by IPC of several million dollars to establish and support a laboratory for Nathaniel and to pay his salary while he worked seemed like a good one.

There was, however, one additional bit of information that Dr. Peters had before her as she considered her recommendation to the Employment Selection Committee. As part of the application process, Nathaniel had submitted a blood sample to determine his genetic profile, as had all other applicants. The profile showed that Nathaniel had the allele for Huntington disease. When asked about that, Nathaniel revealed that he knew nothing about the history of his biological family because he had been adopted as an infant. After thorough genetic counseling about the implications of this news, Nathaniel still wanted the job at IPC.

To have a clearer picture of the impact of this new information on her recommendation, Dr. Peters requested information from the IPC medical director. This report included the following information:

Huntington disease (HD) is an autosomal dominant genetic disorder with an incidence in North America of 1 in 20,000. It is extremely rare in Asians. Individuals who have the allele for HD will, at some point, develop symptoms of the disease; the usual age of onset is between thirty-five and forty-five years. The disorder is characterized by progressive degeneration of nerve cells in the central nervous system. The patient begins to have involuntary jerky or writhing movements of the arms and legs and facial grimacing. Changes in personality, including inappropriate laughter, crying, episodes of anger, memory loss, and bizarre, almost schizophrenic behavior, may precede or follow the movement disorder; the clinical picture is highly variable. The disorder is fatal, with death commonly occurring when the patient is in his or her fifties, and the patient usually enters an almost vegetative state for the last few years of life. Although we cannot predict the precise age of onset of those symptoms, the fact that Dr. Wu has lived to age thirty without any identifiable symptoms means that he has approximately a sixty percent likelihood of onset by age forty. Soon after the onset of symptoms, a person with Huntington disease most likely would be unable to perform safely or productively in a laboratory setting. Medical care for a patient with HD can be extremely costly, requiring long-term care in a hospital or other medical-care facility. Without testing Mrs. Wu, we can predict that their son has a fifty percent chance of having the allele for HD.

Dr. Peters faced a tough dilemma. Should she recommend that IPC hire Nathaniel Wu? On the one hand, she knew that his skills as a scientist fit well with the special research project. He could help IPC develop new products and bring in a potentially large amount of revenue from his work in the laboratory. That would be to the advantage of IPC in the tough and competitive world of pharmaceutical manufacturing. She also knew that the goal of the special research team was to do long-term research, and no one could predict how long it would take to discover new drugs and treatments. She could not be certain how long Nathaniel would remain a productive scientist. IPC was investing large sums of money to support this special research project. Medical and other costs such as disability insurance, once Nathaniel became symptomatic, also weighed heavily as she considered whether to hire Nathaniel. Therefore, Dr. Peters decided to list the reasons for and against hiring Nathaniel for this special IPC research team and to take that information to the IPC Employment Selection Committee.

Procedure and Discussion

1. Write down one reason why IPC *should* hire Nathaniel Wu for the special research team and one reason why IPC *should not* hire him. Be prepared to discuss your reasons with other students.

2. Form small groups as your teacher directs. In your small group, use the worksheet your teacher provides to analyze *The Case of Nathaniel Wu.* Take the position of Intercontinental Pharmaceutical and base your discussion on the information in the case. For the next ten minutes, discuss the case and list as many reasons as possible that Dr. Pe-

ters might have for hiring or not hiring Dr. Wu to work on this special project at IPC. Ask yourselves: Is this fair? Should this be legal? about each reason for or against hiring. Record all the ideas on your own worksheet.

3. Present your lists to the class and add any new ideas to your own worksheet.

4. It is certain Nathaniel will develop Huntington disease. What role does genetic variation play in this case?

5. What specific costs to IPC are at risk?

6. What potential benefits might Nathaniel Wu bring to the company?

7. In what ways is Nathaniel Wu qualified to do the assigned tasks? What unreasonable costs or risks to IPC would hiring Nathaniel Wu involve?

Homework Assignment

Review both lists on your worksheet and put a star beside the three most powerful or compelling reasons in each column. Decide whether IPC should hire Nathaniel Wu. Explain your answer in a brief paragraph that includes what you think are the three strongest reasons in support of your position. Come to the next class prepared to discuss your position and reasons.

PART B: MAKING ARGUMENTS AND ARGUING ETHICAL ISSUES

You will argue whether IPC should hire Nathaniel Wu. A hearing will take place in front of three of your classmates or other people your teacher selects. This three-person group will serve as deci-

sion makers from the IPC Employment Selection Committee. Study the Employment Hearing Questions your teacher provides; the decision makers will use those questions to decide whether IPC will hire Nathaniel.

Class. Form small groups based on whether you answered *yes* or *no* to the question of hiring Nathaniel Wu. For the next ten minutes, work together in your group, using each other's information and the Employment Hearing Questions to prepare an argument that will convince the decision makers of the group's position. Appoint one member of your group to make a brief opening statement (maximum two minutes) outlining your group's argument. At the hearing, the decision makers will allow each group to make an opening statement, and then they will conduct an open discussion.

Decision makers. While the small groups are consulting, work together to compile your questions, review the instructions, and prepare to conduct the hearing as outlined in IPC's Employment Hearing Guidelines.

Decision makers. Conduct the hearing. When the hearing is over, retire to the hallway or elsewhere to decide the outcome. You have five minutes to prepare to explain the reasons for your decision.

Class. While the decision makers are meeting, discuss the most powerful or convincing arguments for each side. Has anyone changed his or her mind? Why?

Decision makers. Present your decision to the class and explain how you arrived at that decision.

The extension activity asks you to consider whether one's genetic profile (or a portion of one's genetic profile) should be public knowledge.

EXTENSION: THE AMERICANS WITH DISABILITIES ACT OF 1990

The Americans with Disabilities Act of 1990 makes it illegal to discriminate against qualified disabled workers. Disability is defined as an individual (1) with a physical or mental impairment substantially limiting at least one major life activity; (2) with a record of disability (such as a cancer survivor); or (3) with the perception of impairment (this likely will include asymptomatic persons who test positive for genetic disorders). The act also bars employers from questioning prospective workers about their past medical history, but permits them to test job applicants for genetic disorders that affect job performance. Employers can require medical exams after a job offer is made, but the results must be kept in a separate confidential medical file available only to the employee's supervisor, providers of emergency treatment, and the government.*

1. According to the wording of the Americans with Disabilities Act, at the present time neither Dr. Peters nor the Employment Selection Committee would have had access to Nathaniel's genetic profile.

---

*From Morelli, T.E., "Protecting Worker Rights," *The Marker*, Spring 1992.

How would that restriction affect Dr. Peter's recommendation to the Employment Selection Committee? How would it affect the decision of the Employment Selection Committee?

2. Given the terms of the act, would *not* hiring Nathaniel Wu be an example of justified or unjustified discrimination?

3. The Equal Employment Opportunities Commission, the federal agency that monitors employment discrimination, has ruled that the ADA protects persons who may develop a genetic disorder but who do not yet show symptoms of the disorder. Do you think that is a sound ruling? Explain your response

APPENDICES

THE DNA MOLECULE

THE WATSON-CRICK MODEL

Heredity is the transmission of genetic information from one generation to the next. This genetic information is coded in a nucleic acid called deoxyribonucleic acid, more commonly called DNA. Information stored in DNA controls all cellular activities and also determines the genetic variations of the cell and the human organism. Figure 1 shows the various functions of DNA.

Genes store the information in a molecular code. At specific times during the life of the individual, cells and organisms use that information. Thus, the genes provide a set of instructions, a genetic program, for an individual's development. Just as in a data bank, stored genetic information can be used again and again. This stored information interacts with many different environmental variables throughout the life of the individual. The interaction of our genes with the environment makes us what we are.

In 1953, J.D. Watson and F.H.C. Crick, working in England with data collected by M.H.F. Wilkins, Rosalind Franklin, and E. Chargaff, proposed a structure for DNA. Using their model, it was possible to see how DNA could act as a gene. The model of DNA structure was such an outstanding contribution to science that Watson, Crick, and Wilkins were awarded a Nobel Prize in 1962.

Watson and Crick proposed that a DNA molecule is a long, twisted, double-stranded structure. Each strand consists of a chain of smaller units, nucleotides. A nucleotide consists of three still

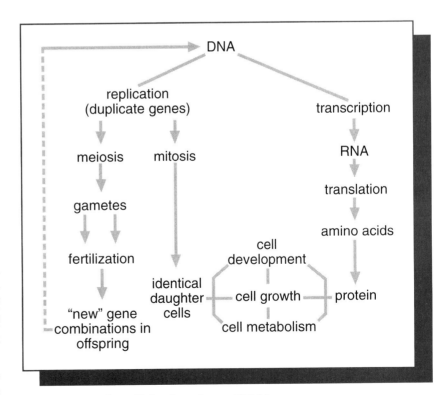

FIGURE 1 ■ The cellular functions of DNA.

139

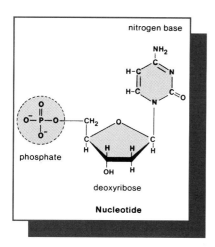

FIGURE 2 ▪ The parts of a nucleotide.

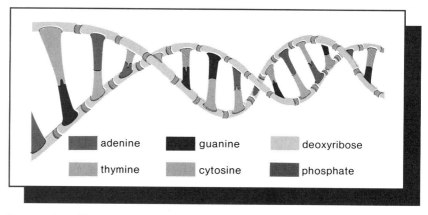

adenine   guanine   deoxyribose

thymine   cytosine   phosphate

FIGURE 3 ▪ Diagram of a small part of a DNA molecule, showing the double helix.

smaller parts: a five-carbon sugar, a phosphate group, and a nitrogen-containing base (Figure 2).

There are four kinds of nucleotides in DNA. The nucleotides are joined in long strands to form the DNA molecule. Each nucleotide has a different base: adenine, thymine, cytosine, or guanine. The sugar-phosphate parts join the nucleotides together and form the backbone of each strand. A base from one strand pairs with a base from the other strand. The two strands are complementary: ad-enine pairs *only* with thymine; cytosine pairs *only* with guanine. The two strands are coiled to form a double helix, much like a ribbon that is wrapped around a pole (Figure 3).

DNA REPLICATION

The Watson-Crick model of DNA made it possible for biologists to explain how all cells in any given person have the same form of DNA, that is, the same base sequences. The chemical bonds that hold the bases together to form the double helix are weak. During cell division, the bonds are broken and two separate strands result. Then a new strand forms on each old strand. The result is two identical double strands of DNA, each of them exactly like the original double-stranded molecule (Figure 4).

DNA relocation explains how the same DNA complement in a fertilized egg will be found in the trillions of cells that make up an individual human organism. ▪

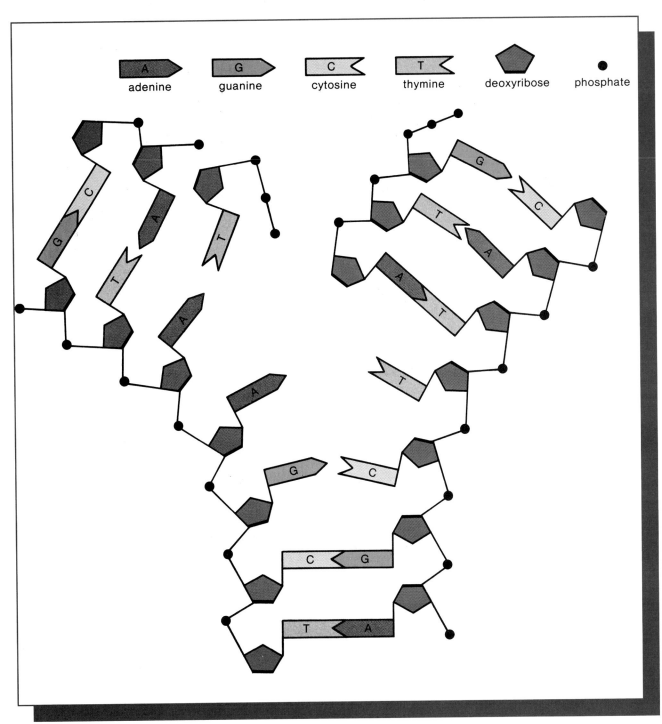

FIGURE 4 ▪ Replication of DNA. The strands come apart at the bonds between the nucleotides. New nucleotides, which temporarily bear extra phosphates, are added one by one. Eventually two new DNA molecules are produced.

THE GENETIC CODE AND PRODUCTION OF PROTEIN

DNA controls all cellular activities and also determines the genetic variations of the cell and the human organism. The DNA inherited by an individual helps to determine the uniqueness of that person's structures and functions by controlling the production of protein. Protein molecules have a tremendous variety of shapes, which allows for extraordinary versatility in the functions they can carry out. An individual's DNA determines which proteins are produced.

Proteins are usually large, complex molecules. They are built from smaller units called amino acids. There are twenty amino acids; their sequence determines the size and shape of a protein. The sequence of amino acids, in turn, is determined by the sequence of nucleotides that make up the DNA molecule.

The chromosomes, with their multitude of genes, are located in the nucleus of the cell. The structure where protein synthesis takes place, the ribosome, is located in the cytoplasm of the cell. That creates problems. DNA is restricted to the nucleus, but it has the code for putting together specific amino acids in a precise order to form a particular protein. The site of protein manufacture is in the cytoplasm, on a ribosome. How can a protein be synthesized on a ribosome when the code is in the nucleus?

The molecule that links the chromosomes in the nucleus to the ribosomes in the cytoplasm is RNA (ribonucleic acid), another kind of nucleic acid. It is very much like DNA, and it also consists of chains of nucleotides. However, the sugar in each RNA nucleotide is ribose instead of deoxyribose. Another difference is that thymine is not present in RNA. It is replaced by a similar molecule, uracil. RNA is synthesized by copying a strand of DNA. It then moves from the nucleus into the cytoplasm, passing through the nuclear membrane.

Three different kinds of RNA occur in cells. One kind (ribosomal RNA, or rRNA) makes up the ribosomes, along with a number of proteins. Another kind is messenger RNA, or mRNA. It carries the DNA message from the nucleus to the ribosomes. A third kind of RNA is transfer RNA, or tRNA. It transfers amino acids from the cytoplasm to the ribosomes, where they are added to a growing chain of amino acids being built into a protein molecule.

Proteins consist of long chains of amino acids. The twenty different kinds of amino acids can be connected in any order. The order is important for protein function, however, so a protein molecule must be built correctly. How do the DNA instructions guarantee that?

Perhaps each base in a DNA strand codes for one amino acid? That will not work, because four bases could code for only four amino acids. A code of two bases together could account for sixteen amino acids—still not enough. Biologists reasoned, and later proved, that a sequence of three bases—a codon—codes for one amino acid. With four different bases, sixty-four groups of three are possible.

Scientists have found out which triplet of three bases codes for each amino acid. We now know that most of the sixty-four triplets code for some amino acid. Some amino acids may be specified by two, four, or even six different codons. Others require a single codon before they will be added to an amino acid chain. The triplets that do not code for an amino acid provide punctuation to the message. They provide the "start" signal for protein production, and they signal when the chain of amino acids is complete. The genetic code is shown in Table 1.

We can summarize the process of protein production as follows:

1. DNA contains the code for protein production.
2. The coded instructions are transferred to RNA when it is synthesized on one of the two DNA strands (Figure 1).
3. The mRNA carries the instructions from the nucleus to the ribosomes.
4. At the ribosome, triplet mRNA nucleotides (codons) specify which amino acids will be brought together in sequence.
5. At the ribosome, the amino acids are attached, one at a time, to the end of the growing protein chain.
6. At the end of the process, a new protein molecule has been formed. Figure 2 illustrates protein production.

Materials

For each team of two students: pop beads: 10 each of black, white, red, green, yellow

Procedure

The protein insulin, a hormone, lowers blood-sugar level and increases the storage of glycogen in muscles and the liver. Insulin is produced in the pancreas. A genetic disorder can interfere with the normal functioning of the pancreas, resulting in a condition known as diabetes mellitus. The insulin protein has two chains; the first six amino acids in one of two chains are:

THR LYS PRO THR TYR PHE.

1. Work with your partner and write down the DNA code for those six amino acids. Use Table 1 to find the code for each amino acid.
2. Use your pop beads as DNA letters (nucleotides) to make a string of DNA for this part of the insulin molecule.

   Key: black = adenine (A)
   white = thymine (T)
   red = cytosine (C)
   green = guanine (G)

   You have constructed a partial model of the gene for insulin.

3. Each team will record on the chalkboard its DNA code for the six amino acids of insulin. Are all the DNA codes the same? Why? Should they be?
4. Suppose that the six amino acids make up the entire insulin molecule. The code for the insulin molecule would then be formed by a very small piece of DNA that is part of a very long DNA segment. How does the cell know where to start and stop decoding the insulin message in this long DNA segment? Examine Table 1 for help to answer that question.

TABLE 1 The genetic code.

First Base	Second Base A or U		Second Base G or C		Second Base T or A		Second Base C or G		Third Base
A or U	AAA *UUU* AAG *UUC* } Phe AAT *UUA* AAC *UUG* } Leu		AGA *UCU* AGG *UCC* AGT *UCA* AGC *UCG* } Ser		ATA *UAU* ATG *UAC* } Tyr ATT *UAA* ATC *UAG* } Stop		ACA *UGU* ACG *UGC* } Cys ACT *UGA*] Stop ACC *UGG* ∫ Trp		A or U G or C T or A C or G
G or C	GAA *CUU* GAG *CUC* GAT *CUA* GAC *CUG* } Leu		GGA *CCU* GGG *CCC* GGT *CCA* GGC *CCG* } Pro		GTA *CAU* GTG *CAC* } His GTT *CAA* GTC *CAG* } Gln		GCA *CGU* GCG *CGC* GCT *CGA* GCC *CGG* } Arg		A or U G or C T or A C or G
T or A	TAA *AUU* TAG *AUC* } Ile TAT *AUA* TAC *AUG* } Met-Start		TGA *ACU* TGG *ACC* TGT *ACA* TGC *ACG* } Thr		TTA *AAU* TTG *AAG* } Asn TTT *AAA* TTC *AAG* } Lys		TCA *AGU* TCG *AGC* } Ser TCT *AGA* TCC *AGG* } Arg		A or U G or C T or A C or G
C or G	CAA *GUU* CAG *GUC* CAT *GUA* CAC *GUG* } Val		CGA *GCU* CGG *GCC* CGT *GCA* CGC *GCG* } Ala		CTA *GAU* CTG *GAC* } Asp CTT *GAA* CTC *GAG* } Glu		CCA *GGU* CCG *GGC* CCT *GGA* CCC *GGG* } Gly		A or U G or C T or A C or G

Note: The DNA codons appear in regular type; the complementary RNA codons are in color. A = adenine, C = cytosine, G = guanine, T = thymine, U = uracil (replaces thymine in RNA). In RNA, adenine is complementary to thymine of DNA; uracil is complementary to adenine of DNA; cytosine is complementary to guanine, and vice versa. 'Stop' = chain termination or 'nonsense' codon. 'Start' = signal to begin protein synthesis. The amino acids are abbreviated as follows:

Ala = alanine	Asp = aspartic acid	Glu = glutamic acid	Ile = isoleucine	Met = methionine	Ser = serine	Tyr = tyrosine
Arg = arginine	Cys = cysteine	Gly = glycine	Leu = leucine	Phe = phenylalanine	Thr = threonine	Val = valine
Asn = asparagine	Gln = glutamine	His = histidine	Lys = lysine	Pro = proline	Trp = tryptophan	

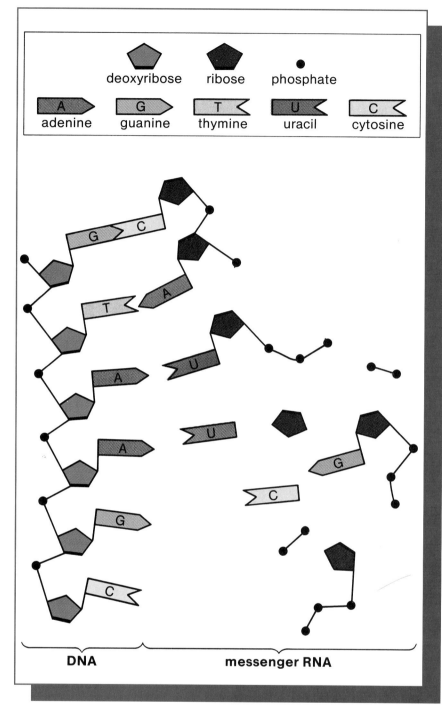

deoxyribose   ribose   phosphate

adenine   guanine   thymine   uracil   cytosine

DNA   messenger RNA

FIGURE 1 ■ Formation of part of a strand of mRNA on one strand of a DNA molecule.

5. Add pop bead "words" to your insulin "gene" so that the cell machinery can find this insulin gene during production of the protein.

6. Remember that DNA remains in the nucleus. What must happen before the genetic code can reach the ribosome, where production of the protein insulin will take place?

7. Use your model of insulin DNA to form the sequence that will be able to move from the nucleus to the ribosome. Use the key in procedure number two, with one exception. What is it?

Questions for Discussion

1. Propose some explanations for the fact that DNA does not move from the nucleus to the ribosome.

2. Assume that the DNA codon GGG mutates to GGT. Would that mutation have any effect on the synthesis of the insulin molecule?

3. Assume that the last DNA codon, AAG, mutates to AAT. Would that mutation have any effect on the synthesis of the insulin molecule?

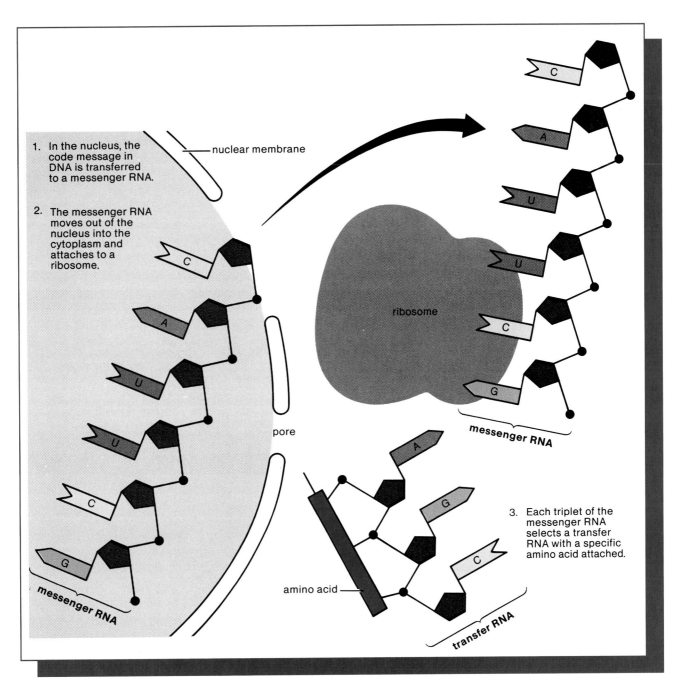

FIGURE 2 ■ How DNA determines the formation of a protein.

4. The ribosome moves along the messenger RNA as it "reads" the code. The amino acids are joined to each other in the order coded. A protein molecule is formed.

5. After delivering its amino acid, transfer RNA can pick up another amino acid molecule.

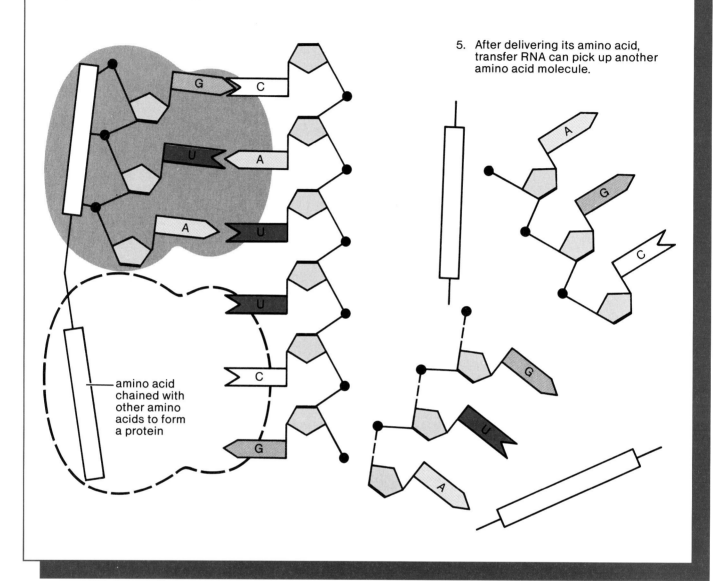

amino acid chained with other amino acids to form a protein

GLOSSARY

Genetics is the study of the nature and sources of biological variation. Natural selection of variations is a major mechanism of biological evolution.

TEN MOST-USED TERMS

allele: one form of a gene

dominant: a variation that can be produced by a single allele, no matter what the partner allele does or does not produce

gene: a sequence of DNA on a chromosome pair that codes for the production of a protein

genotype: the paired alleles that produce a phenotype

heterozygous: having two alleles that are different for a given gene

homozygous: having two alleles that are identical for a given gene

phenotype: the observable characteristics of an organism, produced by the organism's genotype interacting with the environment

recessive: a variation that can be produced only by a homozygous pair of alleles

trait: a characteristic common to all members of a species

variation: the form that a trait may take

OTHER IMPORTANT TERMS

autosome: any chromosome other than X or Y chromosome

chromosome: the structure in the nucleus that carries genetic information in the form of the nucleic acid, DNA

DNA: the molecule containing hereditary information; exists as a double helix composed of two complementary strands of nucleotides

diploid: a full set of genetic materials—two paired sets of chromosomes—one set from each parental set

gametes: mature male or female reproductive cells—sperm or eggs

genome: the total quantity of DNA in a person's somatic cells

haploid: a single set of chromosomes (half the full set of genetic material) present in egg and sperm cells

homologous chromosomes: the members of a chromosome pair that are identical in their visible structure and bear genes of the same trait

hypothesis: a tentative explanation of some phenomenon that science seeks to confirm or disprove through tests

linkage: in inheritance, the association of different genes due to their physical proximity on chromosomes

messenger RNA (mRNA): a nucleic acid molecule produced by transcribing a nucleotide base sequence from DNA into a complementary sequence of RNA

multifactorial: a trait influenced in its expression by many factors, both genetic and environmental

mutation: a change in a gene, or the process by which such a change occurs

nucleotide: the basic unit of nucleic acids (DNA and RNA); composed of a sugar, a phosphate, and a base (adenine, cytosine, guanine, or thymine [uracil])

sex chromosomes: a chromosome (X or Y) that plays a major role in determining the sex of an individual; a chromosome that is not an autosome

theory: not a guess or an approximation, but an extensive explanation developed from well-documented, reproducible sets of experimentally derived data from repeated observations of natural processes

triplet: a sequence of three nucleotides (bases) that specifies one amino acid